Bioclock Farm

J. Fabian Daly

Guy & Lo,
You are a brilliantly creative and challenging couple.
Each morn happy voices are calling life a song for two to love.
All the best.
J. Fabian Daly

Birchfield Press

P.O. Box 573
Accord MA 02018-0573
(781) 749-7563
BioclockScience.com
Bioclocks@hotmail.com

Printed in U.S.A.

I appreciate and thank all who contributed. Friends, teachers, librarians, questioners. Without their help, bioclock science would be years behind its present development.

Staff: Eugene Daly, Editor-in-Chief. Pam Engel, Office Manager. Erin Moore, Medical Researcher. Gerard Whittemore, Advertising Director. File photos by Ryne Kittock and Les Cuneo. Howard Ayen, Sally Basler, Paul Brennan, Sandra Burbul, Gloria Clarke, Fred Curtis, Robert W. & Tim Engel, Dana Gelin, Walter Hartel, Janny Henderson, Richard Fitzsimmons, Tom O'Malley, James "Dean" Moriarty, Walt Power, Mary Walsh, Karen Zeller. The staff thanks Artco, Inc.

Consultants: The Daly, Evarts, Hanneman, Kittock, Pappas and Gandenberger families.

Helpful evaluations of bioclock science were made by Keith Atkinson, Patti Putman-Casey, Robert M. Engel, Bill Marganski and Shawn Clark. The Harvard Medical School and the University of Massachusetts Boston gave me superb and most helpful education.

ISBN 978-0-9702401-1-8

THIS BOOK IS DEDICATED TO GARTH EVARTS, *brilliant farmer, friend, and humorist who was always ready to share his expertise in human nature, plants, and animals with me. His hospitality made* Bioclock Farm *possible. For hundreds of years, people have traveled to the American west for adventure, work, and fulfilling dreams of starting farms of their own. Some have found there are not just corn, wheat, and soybean fields along with livestock pastures, but hard-working, curious, family-loving farmers such as Garth. For many years, Garth's achievements helped to fill "the food basket of the world." My hope is that every reader will feel he or she has met Garth. His life teaches us the dignity of labor and to be positive about any work we do. His life renews our faith in human nature.*

ON MY WAY TO THE HARVEST FIELDS, I met Garth Evarts who had been farming in West Concord, Minnesota for nearly thirty years. If you count the years of his boyhood chores, he had been working on the family farm for nearly fifty years. He was the fifth generation of the Evarts family to farm the same land, an achievement spanning more than a hundred and fifty years. Within the first few minutes of our meeting, Garth had become my guide and was giving me his special brand of education: "There is a Minnesota custom that goes with the changing of ownership of a farm. In that custom, the farmer walks his—or her—farm with the person who wants to buy the property. During the walk, the farmer gives the history, traditions, and reality of farming that land."

Garth began a walk of his farm with me to demonstrate farm tradition. "A farmer is always worrying about the weather. Even when he is only in the crop-planning stage, he's thinking about at least planting some crops that will survive a really dry season and some crops that will survive a really wet season. Then, everyday whether plowing, seeding, fertilizing, cultivating, or harvesting, he's taking the weather into consideration. At all times, he needs to manage finances real carefully." Garth went into a long moment of reflection. "One bad purchase of a big piece of equipment that can't pay for itself can be ruinous."

He stopped at a large, flat-bed trailer and lifted it by the hitch to move it over. He was certainly stronger and sturdier than his average height and weight suggested. He wore glasses and his dark tan contrasted with his light brown hair.

In the center of the farm was a gentle rise of green land trying to be a hill and we stepped to its summit. Garth pointed down and away to his distant pasture where a number of black-and-white cows were grazing, "If the farmer has cows," Garth nodded, "he especially has to make time for them. They must be tended to and milked twice a day, seven days a week."

"One question—many of those cows have both black and white markings—do you consider them to be black cows with white markings or white cows with black markings?"

Garth must have considered this a deep question because he stared at me for a moment. "For the Holsteins"—he made a quick finger-pointing gesture toward his cows—"I'm gonna say white with black markings. Many are more white than black and most have all-white legs."

"Okay. Now, I've heard that once the cows are in the pasture they come home—back to the barn—on their own?"

"Oh, no, someone usually has to go down and bring them in—especially if it's hot weather and they've found a shady spot—occasionally, one will come back and the others will follow her, but they just want to get into a relaxed, undisturbed state—of course, someone taking time everyday to train one for a competition or fair can get surprising results."

Garth spoke about livestock auctions and the art and science of judging farm animals as we went along a tractor path. "The experts begin by evaluating a list of conformational features of the animal." He gave examples. "Many really dedicated owners and trainers spend long hours grooming and training special cows and bulls for voice commands and for maintaining discipline while being handled. Many of the owners are doing it for the sole purpose of winning recognition in organizations and associations such as 4-H and other groups promoting the working farm. Some, I think, do it just for the jawing that always delays the start of a grange meeting."

As we walked, Garth treated all my questions with respect and asked direct questions about my personal interests. "Where is it you want to go with your studies and work?" He spoke in a spontaneous and cheerful way.

I felt most relaxed in talking openly with him: "For, many years—since high school—my innermost goal has been to learn how the human brain functions. I'm discovering if you really want to know the answer to that, you study the four general states of existence everyone goes through everyday: 1) being awake, 2) falling asleep, 3) sleeping and 4) awakening. When you classify the behaviors and events that occur within each of those general states, you notice many of the behaviors profoundly and scientifically are associated with a distinct time of day, either *daylight, dusk, darkness or dawn*, the four phases of the 24-hour earth-sun light cycle.

"Examining those associations for cause and effect, I find the four general states have significant relationships to, and to some degree, are initiated by, the four phases of the earth day. Certainly, the external events and the internal states work together to give us our *(1) daylight response of being awake (2) dusk response of falling asleep, (3) darkness response of being asleep and (4) dawn response of awakening.*"

Garth looked up to the sky as if wondering how the light of day was causing behaviors.

"Now, in bioclock science, daylight, dusk, darkness and dawn are not just the time of day or how much light we're experiencing. They are the durations within our environment that we all experience that have been formed by, and consist of, all the physical forces acting on us and our planet, such as sunlight, gravity, electro-magnetism, rotation of the earth on its axis, atmospheric conditions and geographical location.

"Together, the simultaneous and sequential occurrence of these four external events and our four internal states result in a 24-hour bioclock in everyone. This bioclock can be traced back mathematically, day by day, to the very first 24-hour bioclock. That calculation results in the inescapable conclusion that the organizing principle of human development at every stage consists of four parts. The genome, zygote, embryo, fetus and neonate—all do their sequential functioning based on a series of *four responses* to *daylight, dusk, darkness and dawn.*

"Furthermore, in every mature individual, scientific measurements of behaviors lead to the scientific conclusion that the brain, itself, is organized in *four subdivisions* that function under the influence of *the four responses.*"

Garth drew in his lips as if he was pondering.

"Can I try to convince you that it's a good, fulfilling analysis of science to see the 'light-dark' cycle of the earth day as having four phases? A full sun of daylight does not instantly change over to darkness or darkness instantly change over to daylight, rather, dusk and dawn provide transitional phases."

"Yes," Garth confirmed, "there are transitional phases."

"Well, Garth, if every human being has a 24-hour bioclock consisting of a rotation of four different responses, that bioclock must have four different regions in the brain that coordinate those repeated responses from one day to the next. Most mature individuals fit into the pattern of being awake during daylight, falling asleep at dusk, being asleep during darkness and awakening at dawn. That means there is a four-part, 'awake-asleep' bioclock that utilizes a 24-hour loop of the nervous system. That loop has four sequentially functioning parts, each of which coordinates one of those different levels of consciousness.

"These four parts, *subdivisions* as I prefer to call them, are contiguous regions of the brain, and every day each subdivision has a time of domination over the others that supports a distinct level, or state, of consciousness. Thus, *the daylight subdivision* dominates to coordinate *full consciousness*, *the dusk subdivision* dominates to coordinate *the descending consciousness* that is *drowsiness*, *the darkness subdivision* dominates to coordinate *the partial consciousness* that is *sleep* and *the dawn subdivision* dominates to coordinate *the ascending consciousness* that is *awakening*."

"There's an actual part of the brain that just deals with dawn?"

"Yes, there are a group of events that get us from asleep to awake that are easily identifiable and attributable to just one of the four subdivisions. This, of course, isn't pinpoint perfect science: each subdivision has developed complex levels of functioning and it's not always easy to tell where the functioning of one subdivision ends and the next begins. Influences from other bioclocks are also present. The nervous system also contains a certain amount of flexibility. Obviously, everyone becomes adjusted to a slightly different daily schedule that is no longer perfectly in sync with daylight, dusk, darkness and dawn, but whenever an individual goes from asleep to awake, that behavioral change and its biochemistry traces back to the dawn response."

Garth stared briefly at the horizon: "I've always been fascinated by dawn—especially in summer when there is so much more light before sunrise."

We talked about the variations in the amount of sunlight each season as the earth changed its angle with, and distance from, the sun.

"If you break the seasons down by measuring the durations of the daylight, dusk, darkness and dawn phases each day for one year," I went on, "it would show a pattern in which each phase lengthened or shortened from one day to the next. Changes are produced within the earth's orbit each day and those changes contribute to changes in the four subdivisions of the 24-hour bioclock each day.

"To put the picture together: I'm trying to develop a new method for studying biology that I call *Bioclocks*. It has the *Periodic Table of the Bioclocks* that's based on the *Periodic Table of the Elements* but it lists bioclocks and their properties rather than elements. It maintains that the 24-hour bioclock that's formed each day incorporates all the other bioclocks.

"Bioclocks other than the 24-hour bioclock are important, but they don't exist on their own: they simply account for little changes in the 24-hour bioclock—they adjust its four parts every day. Each of these other bioclocks is defined by the time it takes, usually measured in the number of days, to make its own pattern of variations in the 24-hour bioclock. Thus, everyday for 28 days, the lunar bioclock changes the duration of the four parts of the 24-hour bioclock to become *the 28-day bioclock*. Everyday for 365 days, the seasonal bioclock changes the durations of the four parts to form *the 365-day bioclock*."

"I've heard," Garth stated, "that something unusual happens with the sun's rays during the first moment of sunrise and at the final moment of sunset—that those times produce special color effects like a flash of green—"

"That's interesting. I have been looking at the blue, short-wave radiation and the red, long-wave radiation, and maybe your comment will further my understanding of the phase changes in bioclocks. I will research that and get back to you.

"By the way, it would be most informative to me to get your observations on how bioclocks affect the farm animals and plants. My method is a practical and scientific way of observing and organizing information about any animal or plant. In observing the cow, for example, I would have to investigate the behavioral and biochemical events that a cow went through each 24 hours and then see if there were any patterns formed over many days. I would begin by asking you: Do cows sleep?"

Garth stared downward in concentration and then looked at me: "At night the cows do settle in. I find they are much more relaxed when I play music at night in the barn. Some of them do hunch down on all four legs or kneel and appear to be sleeping— Sometimes, they do appear to be startled—as if they've just been awakened."

Garth went on to explain why a cow should not lie on its side.

Several cows were under a tree at a distant dip in the land and a few were down on all fours. "Did you ever get to like some cows much more than others—sort of like pets?"

"No-oh," he emphatically stated, "but there have been some cows I've disliked much more than others."

A moment later, he was talking about the use of straw for each cow's bed, a cow's 21-day estrus bioclock, her 238-day calving bioclock and an arranged time with the bull. "Oh, if a cow is pregnant, unusual behavior can occur. A pregnant cow can become quite hyperactive and storm around. It wouldn't surprise me if Mrs. O'Leary's cow was pregnant when she went on her rampage and caused a big city to burn down." He started each new story with a smile, as if looking forward to filling each story with good cheer.

He paused, fidgeted with a strap of his blue overalls, and pointed at a nearby road. "That was once the trail that led for miles over to the Hubbell House, the stagecoach way station, where you could go on to connecting destinations. Long before Minnesota became a state in 1858, my relatives watched and heard wagons drawn by oxen making their way west over the prairie. Eventually, one set of wagon wheel grooves got used over and over and that was how the trail started."

"Ah, yes—the famous prairie schooners."

"Prairie scrappers would be more like it. No one who ever jounced from rock to rock, sprung up and down from one bump to the next, pushed and pulled the wagon out of one mud hole after another, ever called that thing a prairie schooner."

We had walked to a distant corner of the farm and as we started back, the silver white dome of the farm's silo reflected the sun and became the focal point of the entire landscape. It was the little lighthouse on the prairie.

Later, we made a trip to the Hubbell House, a straw-colored, symmetrical, three-story stone building made unique by the attachment of an artistic front wooden porch. Garth stared at the building: "It was built in 1854, has a fascinating history, and you can see it's still in excellent physical condition. The same family has beautifully maintained it as a restaurant since the 1950's."

The walk to the building, the steps and sounds on the wooden portico and the quaint country murals on the inside walls brought me into 1850 and the Old West. We began a delicious meal. "Is this corn," I asked, holding up an ear, "from the nearby field we just passed on the way in here?"

"Oh-o," Garth gasped—"We would never just ask a restaurant if their corn is from a nearby field," he forcefully instructed—"There's a major distinction between 'sweet corn' and 'field corn' and that nearby field may be field corn."

For many minutes he differentiated the major varieties of corn and their uses.

I nodded.

"Garth," I inquired, "that one man—every time he comes into the room, all the employees come over, make a huddle, and he gives instructions while they listen and then they go back to work. It's as if he's a quarterback giving out the play."

"That's Paul Pappas, the owner. We've signaled greetings to each other and he should stop for a visit—I'll introduce him to you."

I was mystified. "What kind of signals did you give each other?"

"Minnesota people have developed the fastest and the most efficient greeting anywhere." He moved his hand up in front of his chest, slightly uncurled his fingers outward in a flick, and then lowered his hand. "When two farmers are driving their tractors past each other, there's only time to lift the fingers above the steering wheel for a quick wave." He again did an almost unnoticeable finger flick.

"Good thing I didn't blink."

A moment went by and Paul did stride over and sit. Garth introduced us.

They quickly went on to what seemed a back-and-forth routine of one of them naming a member of the other's family and asking how that individual was doing. Sometimes one would only speak a name in a questioning tone. They began talking about the recent date of one of Garth's daughters, Sue, with one of Paul's sons, Paul, and agreed it had occurred without incident.

The conversation changed over to exchanges of Garth talking about crops and bushels per acre and Paul speaking about breeds of animals and the restaurant business.

I listened carefully in hopes of learning this farmese language. In a relaxed manner, Garth frequently checked on my understanding of the discussions—he was making attempts to educate me and, out of courtesy, getting me to join the conversation: "Jay, you do know the difference between corn-fed beef and range-fed beef?"

I remembered reading something about iron and the range in Minnesota and proudly concluded: "The range-fed beef contains far more iron."

Each of the men gave me a philosophical look, each made an uhm sound, and then they went back to their conversation. Suddenly, Paul was up and became invisible.

Garth and I continued socializing in the midst of which he was introducing me to employees entering our area, several of whom turned out to be members of the Pappas family. He spoke to the diners at one table after another and always included an introduction of me as a visitor from Boston.

I noticed Garth gave a warm smile and greeting to anyone who came anywhere near our table, including several people he had not met before. He really was there to joke and talk with people. "Right now, I would like to hear more about bioclocks—" He was a delightfully curious person.

"I mentioned *The Periodic Table of the Bioclocks*, my summary chart that's used to study each species by its bioclocks. The Table's organized around the theme: '*Follow the Fours*'. That is, *the four phases*, *daylight*, *dusk*, *darkness and dawn*, lead to *four consecutive responses* in every species and *the four responses* require *four contiguous subdivisions*.

"My hypothesis is that the 24-hour bioclock in the human is virtually a self-evident principle of organization.

"Since a bioclock is a biological or behavioral pattern, observations are used to study each subdivision for the events occurring during it and see if they repeat.

"My conclusion is that the universal experience for the human individual is that everyday everyone experiences being awake, going to sleep, sleeping and awakening. This is based upon the fact that *the four responses* have led to the human brain functioning in a 24-hour behavioral bioclock with *four subdivisions*. The Table uses our species, Homo sapiens, as a working model with which to compare species.

"To make the best use of my method, you have to remember

that the human no longer functions in perfect sync with daylight, dusk, darkness and dawn and therefore the duration of time an individual is awake, drowsy, asleep or awakening on a given day will routinely be different than the hours for the earth-sun light cycle on that day. For example, on June 21 in Boston Massachusetts there's about 15 hours of daylight, 1 hour of dusk, 6 hours of darkness and 2 hours of dawn. That's a 15+1+6+2 total. However, if an individual in Boston on that day is awake for 17 hours, drowsy for half-an-hour, asleep for 6 hours and awakening for half-an-hour, that individual's durations for his or her 24-hour bioclock are 17+.5+6+.5.

"Each day, the four durations of the 24-hour bioclock occur because each *subdivision*, or region of the brain, rotates into a time of dominance over the other *subdivisions* to produce its unique state of behavior. Each day, the four durations are noted and referenced as *the four periodic numbers for that specific day*. A log is kept of *periodic numbers*.

"Every species, then, has an overall four-part bioclock and any other bioclocks within it work by making changes in the durations of those four parts.

"My theory begins with evidence that the genome of the first cell had four sequentially functioning parts, achieved replication and divided into two cells. And the biochemistry forming those cells was influenced by the forces and conditions behind daylight, dusk, darkness and dawn.

"In my hypothesis, then, every cell has a bioclock of its own and when two or more cells join together, they also form an overall, group bioclock. Thus, when a group of cells does function together as a unit, each cell is an interacting bioclock with the overall bioclock—each cell is really a team member important for its contribution to that overall bioclock.

"Once you know that everyday there is a change in the length of time each subdivision of the overall bioclock is functioning, many days of its *periodic numbers* can be observed and logged to see what, if any, unique patterns are formed. Then, in extensive observations over days or weeks, any developmental stage could be studied as a species-wide bioclock, such as the embryonic bioclock or the fetal bioclock.

"To summarize, the Table is a research and learning tool that can be used to match a subdivision with the chronology of events related to it, thus map the anatomy and physiology of behaviors. To that end, the study of bioclocks advances brain research by showing that the nervous system is a hierarchy of neural interactions and routes.

"I recommend an individual log the durations of his or her four responses on a graph each day for months to see if any patterns are formed over and above the 24-hour pattern. Every pattern found constitutes important self-knowledge. If a bioclock is found, the total length of time it took to form its pattern is that bioclock's *periodicity*—which is usually measured in days—and that number is used to label it as a bioclock.

"In this way, the human 24-hour bioclock has been shown to contain other bioclocks that have been identified by their periodicities: the human 24-hour bioclock is shown to be composed of *the 28-day lunar bioclock* that makes small variations within the 24-hour bioclock each day for 28 days, *the 365-day seasonal bioclock* that makes very small variations within the 24-hour bioclock each day for 365 days, and *the reproductive bioclock* that makes extremely small variations within the 24-hour bioclock each day for years.

"Some species also achieve a sequential interaction of the subdivisions in milliseconds. An animal with a heartbeat, for example, experiences a surge of blood to the brain that can be identified for its effects on the subdivisions. The cow, as an example, can achieve a distinct behavior within the milliseconds after a heartbeat. Many heartbeats can produce an observable pattern of brain dynamics that constitutes a bioclock. So, ah, anything from a gaze to a neuromuscular movement made by the cow may actually require under or over a millisecond but each distinct event can be analyzed as a fundamental unit of behavior.

"The *heartbeat bioclock* is defined as: 'A sequential interaction of the subdivisions following the heartbeat and/or the use of neurotransmitters that results in a behavior.' That definition usually saves the need for precise measurements of every behavior. For example, measurements could be made of the gravitational effects of the sun and moon on the neuronal interactions contributing to a behavior, but often all that's needed is to see that the individual's behaviors form a pattern during the lunar bioclock."

Garth gave me a quizzical look: "I don't follow everything that you're getting at, but let me ask: "How does the farmer benefit from this new knowledge?"

"First and foremost, bioclock science finally explains how the brain functions and how to make the best of that information. That means everyone can now use more of his or her brain power by understanding and better utilizing brain resources. For example, each individual can see how sleep and dreams are part of the process of forming and using memory and controlling mind-body interactions. For the first time, we now have a model of the functioning brain that lets us know the general neural routes of pain. And that allows us, sometimes, to direct our thinking and physical behavior to lessen pain.

"Certainly, a major benefit is for those involved in neuromuscular disorders. This science gives us the major routes the brain uses for information processing. We now better understand how and why a process is occurring. That advances research in brain disorders. We now know that the control and attention center resides in the left hemisphere—with some qualification for the left-handed individual—and from that we can follow the regions used in coordinating neuromuscular and memory functions. That gives us a major, new understanding of where and when neural malfunctions result in autism, Alzheimer's, and many other disorders."

Talking to Garth had greatly helped me to organize my thoughts. "I'm just now editing more than ten thousand pages of my notes into an updated draft and when that's completed, I'll have a detailed discussion with you on how to benefit from this new knowledge."

ALMOST TWO YEARS LATER, I was on my way to a second trip to the harvest fields. The first trip had been an entertaining and educational experience and I was pleased to return. That trip taught me that a farm is an excellent place for observing the bioclocks of many plants and animals and how to use that knowledge for better appreciation of life. It was natural, then, that on this trip I planned to visit the Evarts farm.

When I reached Dodge County, I phoned.

"Evar-ettes," Betty, Garth's wife, answered in a clear, bright, proud voice. She always answered the phone with the family name and that was part of her wonderful spontaneous humility, joyful family spirit, and impressive team work. She taught everyone that a call to her at the farm not only reached her as the person who was answering the phone, but it was also a call that reached a family.

When Garth came to the phone, I greeted him and asked how everything was going.

"The figures for last year are out," he brightly announced,

"and Dodge County was again in the top three counties in the United States for bushels of corn per acre."

"Congratulations, you must be proud."

"Those figures always make me feel humble. Lancaster County was in the top five again and those Amish farmers do it without the use of motorized vehicles."

On this visit to the farm, I determined that the next morning would be a good morning for new adventures. I got up at 4:00 a.m. This was an experience that was allowing me to observe for one morning what most farmers observe every day.

As I stepped outside, I looked down. Some crickets were springing hectically about as if organizing into a marching band that was using little drums and fifes to crack and phew a grinding, rockslide rasping to the ear. Are they creating a thousand-ring circus? Joining in a mass dance-and-sing-along? Rioting? What is going on?

Suddenly, somehow, their chant synchronized and "Creep, creep, creep" became constant. I wondered if there was one greatly admired cricket whose name was "Creep" and all the others were shouting his, or her, name—"Creep, Creep, Creep."

I wondered if maybe there was a special exercise being done by the crickets and they were yelling to each other to "Creep, creep, creep."

I wondered if one of the crickets had socially unacceptable habits and all the others disliked him or her and were hollering out, "Creep. Creep. Creep—"

I looked up. Innumerable tiny lights in the sky drew me into a magic theater of vastness. The stars, planets, and moon gave me an unbelievable sense of fulfillment. This is my own planetarium, I happily concluded. A cosmos-wide black web with small silver shining pearls randomly placed on its strands—Actually, the black drape of sky is dark bluish-black at the western horizon, medium-black above, and ever so slowly easing into the hint of a pink layer at the eastern horizon—Our solar system introducing its featured act of the morning, not with a drum-roll but with a light-roll.

For a long time, I gazed at the reddening pink easing upward and outward as it rose over the black land, the floor-lighting of an enormous stage preparing to present the star of the day –

A breeze set in as the cooler air moved toward the warming air.

A bird made the sound "Tweep." That is the first sound of the first bird to start the morning singing. I looked about, expecting something to be next.

"Twee, tweep, twees, tweets, twills, twillups and tweeeeeeeeps," became chattering and arguing intonations as the birds went into high-pitched rings and low-pitched rings. What-was-orchestra converged with what-was-chorus and racing vocalizations took over as one shrieking, hectic chicken "twip-twur" was cut off by another chicken "twip-twur." Each indignant chicken was going at it with a louder and more shrieky, frantic twip-twur.

Other critters with wild and way-wild slangs were scurrying about in the brush but they were contestants not seen—they sent my imagination to the mysteries of a rain forest and then my reality to the intense sounds of a farm forest.

I looked up to the tops of the tallest trees and followed the descending flight of branches. Somehow, the branches became wider and longer as they were slowly descending around the dark brown tree trunks and touching-down on low-lying brush-covered runways.

My attention kept alternating from a stare at a *specific branch* to a gaze at *branches*. How was a view of *a single branch* changing to a view of *intermingled tree parts*? How did I concentrate on one branch only to find myself in a thoughtless gaze at a giant nest of entangled branches? It must be that the control center for the dominant eye in the left brain is alternating its single isolated image with the full visual-field picture from the other eye's visual-field in the right brain, a physiology resulting in the alternating pictures in the visual mind. It is the process that allows either *the single tree* or *the group of trees* to be experienced.

Amazingly, the experience of seeing branches meshed and merged in one scene results in one amazingly unified, relaxing experience.

Beyond these nearby trees and shrubs, the background was green-rows of corn delineated by black soil-rows, a geometrical latticework extending to the sky far away. I walked along in baby steps, counting to three to keep count with the pattern of the field: it was two steps along one corn plant followed by one step across a dark line of soil. After many counts, my entire attention was taken over by the frickle, frapple, and fripping of corn plant leafs doing a war dance to the beat of whipping, whacking and whopping winds.

I stopped walking and wondered: Why is it that when I drive along lines of row crops, the close part of the rows rapidly march by to keep up with me, but the distant parts of these rows seem to be staying in place?

Dawn was decorating the farm so beautifully that I vowed to enjoy the experience more often. I felt my nose move as I breathed in and the breath became long and deep without my so directing it and I went into a moment-before-sleep relaxation.

The memory came into my mind of that evening, years before, when I had stood near this spot and had seen a deer easing along one of the soil-rows of a cornfield. I looked down the spaces of these rows for the deer, but did not see it. I felt happy: "You, at least," I said out loud to myself, "did have a deer sighting and it's now an enchanting memory." I closed my eyes to savor the moment and kept them closed in hopes a mind picture would occur but somehow the deer image would not reappear. "You really enjoyed your first visit to the farm."

Though that trip was some years ago, somehow the relaxed mind sweetens a clear view of a friendly farm vista. It is an emotional experience to have a fixed-in-the-mind-forever farm image and feel this land as timeless. The emotion may be called "sentimentality" by some who cannot let themselves be vulnerable to a simple, spontaneous display of a childlike feeling of being happy. I turned away, yet I glanced back and that filled me with a feeling of love-from-the-heart for this instant: this review of this time and place formed into a sense of meaningfulness as enriching as any view I had had anywhere. I felt renewed but could not decide if I should get to the work I needed to get to or continue to relax and let my mind drift into past events. A fast breath-out sent the mood away.

I looked out to the black road that meandered away for many miles to the town center—the memory of a drive to that center crystallized in my mind's eye—the steel strands of railroad tracks leading to the small town center with the huge grain elevator—above which the cloudy dark sky had gone into white-silver flings of lightning—then the explosion of thunder—a primordial "Ha" that had arrived so full of fury, energy, and loudness that I had breathed in and heartbeated into an adrenaline all-attention high—and forever learned that thunder in the city was not the whack-and-crush-the-chest thunder o'er the prairie.

Shutting my eyes, my mind filled with the scenes and splats of that electrical storm: Hail upon hail so thick and so quick that it is glass droplets chill-plattering against the ground—an instant by instant series of causes and effects in which each shatter stays just long enough to be identified—a video game that mesmerizes the eyes and sets a trance –

Opening my eyes, I looked down at my clothes and was astounded that they were not soaking wet from the rain. I looked up, scanning for the clouds of weather: "Clear and dry—a bright morning of early sun—I've virtually re-lived—How can these memories be so intense that they can make me believe that I was actually experiencing that rainstorm again?"

Garth likes to take the family children outside during distant lightning and thunder. The picture-memory is with me of walking to the barn with him during a storm. "He wants to teach all of us not to be afraid."

I stared out over the land of farms and it riveted my attention to curiosity. It's as if I've been visiting a gigantic museum that featured a live farm as part of its art gallery.

I remembered Elsie, Garth's mother, taking me to a monument dedicated to the Native Americans that is located near the family farm. She had explained that Native Americans existed in this region before the start of recorded farms. "Now, of course, there is the place in North or South America," she had summarized in a tone that would have made anyone believe she was actually a bright, friendly, devoted history teacher, "where the oldest farm existed. That may have been the farm on which a Native American worked with plants and seeds to hybridize their corn. Some say the farm was in Mexico or Peru, but the Evarts farm meets all the requirements for growing corn. So there—that means our farm could have been the actual land on which the first farm existed."

I found myself walking through a field of foot-high soybean plants and stepped out onto the road. I was at the edge of the Evarts farm. I stood and just let the ongoing event of being surrounded and immersed in a farm ease into my mind, fill me with meaning and then ease out of the mind. By summary, it had been a farm mood set on me as a storm-eye of meditation: a pleasant, thoughtless moment of dominance by the right brain.

I turned around and looked up to the Evarts farm for a very long time. At its center, the gentle rise of green land succeeded in being a hill. It had a crown that consisted of a dozen green trees, a two-story white home with a sloped roof of dark shingles, a large barn, and tall over everything, a cylindrical grey concrete silo topped with a silver cap.

The barn's huge dark roof with two slanting sides took over my attention—a gigantic, half-open book has been set over the building.

"Try to imprint this view," I commanded myself, "so that you will always be able to remember it."

My attention went to the far end of a distant gravel road where a series of little, mischievous clouds of dust seemed to be chasing after a pick-up truck. It was all that was moving in the landscape and timescape.

"Who-wish-ohm," came from a puff of the wind's breath. I thought of the seashell held to the ear to furnish flowing fluffs of muffled sounds and I wanted to find a farmshell that when held to the ear would amplify the wandering farm wind's whispering and echoing "Hel-low-oh—"

I took a long breath. Looking out over these vast empires of green is meant to become an indelible moment—a memory that can always be recalled and a memory that I look forward to recalling. Let me roll up this view as if it were a poster that can be taken to a world-class museum's exhibition of the world's best views and there earn an equal award in competition with any other great scene of the world.

The next day, a group of us were discussing the beauty and attraction of house plants. "Myself, I'm into aroma therapy—" Garth announced while smiling, "Betty's rose lipstick really sends me—"

"Oh, Garth—" Betty half-chided, half-laughed.

I found myself in a series of reflex jaw movements and mumbling laugh-grins as I enjoyed the ease with which Garth appeared and instantly had everyone bright, cheerful and gabbing away. He was always lightheartedly orchestrating his delighted fans.

"Oh, you must come outside," Betty excitedly directed us, "and see Garth's gladiolas. They are just the most beautiful flowers I've ever seen."

We went out and admired the flowers, quickly becoming fixated on the two-foot high pastel sprays of reds, whites, pinks and oranges formed by ruffled stalagmitic gladiolas.

When my gladiola trance broke, I found myself asking Betty about the local use of medicinal plants.

"We've been under the medical direction of the Mayo Clinic for so long now that we seldom hear of anyone planting and gathering plants for *the illnesses* or conditions. There are some we hear about who gather various plants to cure each and every thing that ails a body but we don't pay much attention." Betty made a face that seemed to be remembering a terrible taste. "Some plants, I believe, should be grown for nourishing the soul rather than for would-be cures for maladies of the body. Plants that produce colorful flowers are grown for their beauty and the simple delight—the inspiration they give."

I decided not to ask more about medical recipes. "I can't add much to that excellent summary, Betty. Bioclock science does teach that we are genetically predisposed to benefit from colors and the eyes need to be stimulated by colors to be healthy and perform mature functions. In general, when we experience beautiful colors, we experience our brains producing hormones that give us a sense of well-being." I thought of the color red but decided not to bring it up as an exception because some people sometimes responded to that color with heated comments.

“Oh, the Andrews are here to visit—” Betty stated as a car carrying surprise visitors arrived. “I must go inside and make coffee for them.”

As if in a track event, she sprung into a dash. She normally seemed to move at twice the speed of the average person, but there were times when she just simply was in one place and instantly reappeared in another place. I had the belief that each morning when she raced around getting the radio, the washing machine, the stove, the vacuum cleaner, and other units, all going simultaneously, that it was her commotion that woke up the birds, the bees, the poultry, cows and the other farm life.

I marveled at how she took in stride the sudden appearance of five or ten visitors. ‘Coffee’, of course, also meant a hot homemade soup, sandwiches with home-baked bread and cookies.

As the visitors left, Sue, one of Garth and Betty’s four daughters, stayed outside with me while the others went into the house.

She stood, arms folded in front and facing up at the sky. Sue was in restaurant management and liked a quiet moment to recover from a day in the hectic public life.

“I believe that when you experience the color green, Sue, and that means everyone, you are receiving a special nutrient for the eyes. In spring, does not the quick appearance of green renew the spirit and energize the attitude?”

Sue replied with a thoughtful, confirming smile.

“During the summer, are we not responding,” I asked, “after, perhaps, the blue of sky, much more to green than to any other wave length of the color spectrum?”

Sue gave a nod and an eye click of agreement.

“Indeed,” I continued, “we have a natural system of values that makes us anticipate, appreciate, and enjoy the environment’s many shades and textures of green.”

"We who are the farm family, of course, have one experience with green that is over and above all the other experiences—" Her voice was deep and vibrant but she was speaking slowly—yet there was an ironic intensity to her words. "Each member of the family gets to watch the winds make waves over the ocean of green fields."

I had the thought that she was speaking pure family truth.

I could see how someone could compare these whirls from summer winds in the farm fields to the ocean: the movements were waves creating their rising and falling in rhythmic sweeps out to, and seemingly over, the top of the distant hill—It was, in the waves of grass swaying out, leaning down, and then springing up in the next wave's swaying, an alluring deep sea on the go.

What the scene meant most to me was that the wind was dancing with green plants. This swirl of air sprung onto the dance floor, took and dipped a plant, held her in place for a moment, lifted her back up, and then fluently swung on to another partner. Absolutely, wind and plants playing, hugging and caressing in a survival dance, indeed, a 'thrival' dance.

"Minnesota need not envy the beauty of any scenery elsewhere," I summarized.

"You're correctly getting to the point," she said, brusquely and surprisingly fiercely—"it's experiencing the mountains and the ocean at the same time."

She seemed now to be bringing a rather seething, personal mood into these events. I had the thought that the scenery was in some way parading in front of her and she was reviewing it—an overly demanding general in a ceremonial review with the troops but someone who was driven to be on guard because of an emotional overload from a past experience.

I quietly and slowly eased ten feet from her side, believing it just might be time to let her find composure.

I began to think about an actual ocean coming right up to a bank of mountains.

Sue unfolded her arms and slowly walked away. She went with little joyful sway-right sway-left head moves as if keeping time to a song—and with dance steps. A contented march into her landscape. She was wearing a loose, white, pleated skirt and it pinched tight behind her from a wind swoop.

I let her be. She seemed to be finding inner peace and joy by easing through a green field and joining the plants in what seemed to be a ritual ceremony of praise and thanks for the feelings of warmth and wellness received from the Minnesota summer sun.

I walked for a long time while thinking about the state's undulating landscape. There had been a time when I thought it was mere flat land.

I found myself standing still but imagining myself in flight, looking down and traveling over the State's varied and scenic topology.

I returned to the present and let my eyes gaze on the open vista. Indeed, Minnesota grasses are so special that everyday wisdoms about their qualities have come into existence, such as: "No matter which of the bordering states to Minnesota you are standing in, Minnesota grasses are always greener than those at your foot."

Another of those everyday super insights deals with the special aromatic treat given to those who are nearby when a Minnesota grass is cut: "Always take the time to stop and smell the new mown Minnesota grass." Indeed, so many tourists have visited the state for these special aromas that bottling companies now supply units filled with aromas such as Minnesota rye or Minnesota barley and hops.

My mood was relaxed and intense at the same time. This was July and I thought back to June and estimated that many fewer shades of greens existed then than now. I made estimates of the number of greens going back through May, April, March, February and January, and that formed a pattern of diminishing numbers. "The 28-day bioclock of a prairie species can be studied individually or in the context of a landscape's 28-day bioclock."

I realized that I was now getting a sense of power out of talking out loud to myself.

Seeing a wild onion at the side of the road, I stepped to it, lifted it from the ground, took my jackknife and made a cut into the bulb. A very pungent onion odor occurred. I touched a finger to the bulb and gave a sample to the tongue tip.

I opted not to taste more raw wild onion.

I wondered if some of these wild onions found in this region had naturally progressed to sweeter varieties or if the onion had needed the farmer to breed it for sweetness. My eyes were filling with onion tears. "From this acrimonious smell and bitter taste, the sweet, wet, crunchy bites with pleasant aroma have developed." I had a blurry gaze at a cloud: "Onion is surely unique and the most crispy of all bites." I imagined the crispiest onion and taking a pretend-bite of it, my mouth watered. "It's near addictive."

Stepping away, I felt my eyes clearing as memory after memory allowed me to roam slowly back over this region's history and paint inner pictures of it. My thoughts had an *in slow motion* quality to them as I stopped and stared at a nearby apple tree from which I had picked fruit that Betty had made into the most delicious apple pies ever. "That pie aroma easing out over the yard is spectacular. So too the memory.

"To what degree is the farm family's love of cooking generated by their own harvesting of food from the land? What a profound, life-changing experience it is to leave the supermarket and go to live only on food grown and harvested directly from the land." I now felt it was perfectly natural to be isolated and talk out loud to oneself.

Observing distant and randomly-located clumps of trees, I began to study each clump, imagining chronological changes in the land that would have occurred with its formation. How many of these groups have resulted from geological formations of sinkholes and how many of them have just simply been planted?

To what degree did a location of edible grasses and food trees promote the first farm? Did the first farm family find a cave or natural shelter near these plants and then begin to farm?

When the Native Americans had begun that first farm in this region, could it have been started as a winter shelter that turned into a year-round home? Certainly, if they had experienced the warmer weather to the south, they would probably have had a practical incentive to locate the first farm of northern America down there. Regardless, it would have been most beneficial for that first family of farming to have begun a tradition of friendliness as part of their acquired learning. The Evarts family farm had a tradition that went back to the Nineteenth Century and any farm predating their farm would have needed such a tradition. For a family to have thrived, there simply must have been a spirit of working together, sharing, and joining in a community spirit of helping neighbors and making visitors welcome.

Most people who came to the earliest farms, it would seem, would only have had to give a friendly greeting and then they would have been met by a residing farm family member who was ready to offer hospitality, including some information on the history and stories related to the farm. Isn't making friends more of a survival factor than making enemies?

Residents and visitors to any farm in this area would have felt a special sense of time and tradition. They would've had this vast, open view to look out over and become filled with confidence and immersed in a sense of fulfillment. They probably would have concluded that the first farm was a home and garden rising out of the primordial plains and beginning the age when survival and thriving meant living in near-perfect sync with the sun by getting up at dawn, working by daylight, preparing for rest during dusk, and sleeping during darkness.

I must take the time to reflect on generations of farm animals being fed on a fixed schedule each day and estimate how long it would take to have a species internalize that event and make it an instinctive behavior that formed a bioclock. *The Periodic Table of the Bioclocks* could be very valuable in such a study as it offers a method to observe neuromuscular and cognitive events and classify each into one of the five categories of heartbeat bioclocks.

Plants known to have been utilized from the earliest farms onward could also be studied. Plants that are being changed from a nomadic existence to a formal schedule begin new biology that could also be examined by bioclock science.

I told many in the farm community that I was piecing together the story of the first farm and the early farms and told them to feel free to make comments. I said the first farmers must be admired for their contributions to the art and science of living. The community gave me points to ponder as well as an oral history of local farm life.

Down through the years, the descendants of that first farm became aware that the tallest plants were almost always the most productive and therefore the most pleasing. Those farmers had found it informative that most of the plants that were exposed to the first rays of the sun became the tallest, the largest, and gave the most rewards. To some degree, those plants were certainly getting their advantage by shading out smaller plants. These taller plants were then given special attention. They were the first to be watered, fertilized and cultivated. The farmer, thereafter, knew the seeds from those plants were the producers of the most rewarding crop and used them for the next year's planting.

Over many generations, these special plants were consumed and both the farmers and their animals grew to astounding heights and strengths when compared to other people and distant animals. This development was probably the basis of the story that Minnesota corn grew to the height of the farmer's shoulder by the Fourth of July instead of just growing to knee high, the way it did in other neighboring states. Indeed, the legends of giant people in the north of Minnesota who could cross the Mississippi with one step probably came into existence from the bragging done about the great growth of corn in Minnesota. Not that it's all myth: how about those bones that were found in the north of Minnesota that were thought to be those of a dinosaur, Tyrannosaura regina, until enough of the bones were found and assembled and it turned out to be a giant chicken? Legends like Paul Bunyan and his ox may also have had their start under the star of Minnesota. Of course, what is factual in Minnesota is often evaluated in other states and declared to be folklore.

I smiled as I realized how often I was mixing real farm and imaginative farm. A learning experience that is both educational and entertaining is the very best. Regardless, for me that still requires I set down a bioclock farm. I stepped back into the soy bean plants, reached down, snapped off a pod and began to examine it. Useful, healthy, disease-resistant plants are often the results of great care in growing and breeding. Other qualities of plants also cause them to be admired and promoted, such as yield, ability to survive drought and/or voluminous water. And important is use in medicinal practices—as with the willow trees and lily family.

I've heard much about soy products, but I really knew little of the plant. I began comparing the pod to that of the pea and the string bean. It seemed strange to me that the soy bean and string bean had ingredients that distinctly promoted human cell growth while the pea seemed to have the best theoretical ingredients for stopping excess cell growth such as the tumor.

THE SEASON HAS SWITCHED from rhubarb, lettuce and peas to apples, corn and tomatoes and I'm having more frequent discussions with "the feisty one" who is idea rich and ready to talk about life's controversies. Bonnie, Garth and Betty's youngest daughter, is an excellent elementary school teacher and is evaluated by everyone as a superb debater. At least, everyone says she's always ready for a debate.

"A close examination of the theme of *The proper manners and decorum a farm child needs for good learning* had taken us back to the earliest farmers and their use of a 24-hour bioclock," I stated to remind Bonnie of where our conversation had left off a few days ago.

I was trying to find a conclusion from bioclock science that agreed with Bonnie's position on the relationships of personal beliefs to behaviors: "The study of bioclocks began when an individual observed rhythms of the sun, moon, seasonal variations, and weather patterns and then consciously responded to them."

"All right," she said acceptingly, "there were observations that led to the establishment of the times for planting, growing and harvesting—but I'm not sure how this subject goes to the heart of the problem—"

I lifted my hand to request a moment to get to my point. "Farmers finding and using a pattern that predicted future farm events often had success with crops. Those farmers of thousands of years ago also improved the use of their minds and bodies by then creating calendars and using even more planning. Still, those farmers did sometimes have a successful crop and sometimes they didn't, and that made it natural for those farmers to seek out and worship the deity associated with the successful crop. My question to you is: As you grew up on the farm, to what degree was your religious experience tied to praying for a successful harvest?"

Bonnie looked up from the book she was reading. "If you're asking me what it was personally like to be brought up in a religious family, I think for me, faith came out of the sharing we did. Sharing was a way of honoring God for what we had been given. Of course, everyone had some individual chores to do in order to learn responsibilities and make the farm successful, but that was just earning the rewards of contributing. Sharing was what made us do things together and we all wanted to do things together. It was sharing that really made us a family and guided our actions each day."

Bonnie set down her book and began to stare me down. "Yes, we did pray for successful harvests and good health and gave thanks for the goods received from God's bounty but all that was really important was to share. *Sharing* was the good book's teaching on how best to serve our faith. God shared his love with us and our sharing was love of God."

"*Sharing* was the basic religious belief of your family and practicing it led to faith in God," I summarized.

"Well, it does not matter which comes first—" Bonnie was emphasizing her words by waving her right hand back and forth, forefinger pointing out—"Sharing leads to love of God and love from God leads to sharing." Bonnie had emphasized each half of her sentence by a forward slash of her forefinger. She stopped and locked together the fingers of both her hands.

I was amazed at how often Bonnie summarized and it was a summary that I would have been pleased to have made in those exact words.

I nodded. "I believe that faith in oneself leads to a religious faith—"

"I would say 'belief in oneself'—" Bonnie firmly suggested.

Bonnie was always considerate in listening to my presentations. She had a special way of interjecting a word or phrase that she wanted me to confirm or reject as a term that held the same meaning for each of us. This special level of communication did not constitute an interruption and made me always try not to interrupt her when she spoke.

"I'd like, for the moment, Bonnie, to explore the possibility that 'faith in oneself' is related to an individual's religious faith. The First Law of Biology explains that every species is organized in four sequentially functioning subdivisions and the Extension for Homo Sapiens explains each subdivision rotates into a time of dominance over the others.

"From the Extension, we know that when the individual is in the awake state, he or she has the use of a hierarchy of neural interactions to coordinate each individual behavior. This *hierarchy* consists of the five *heartbeat bioclocks*, the instinct, neuromuscular, curiosity, reflection and monitoring. This hierarchy is what the individual has available to achieve his or her many levels of complex functioning.

"Now, every healthy individual matures by taking it for granted that his or her own nervous system will work properly. Each one of us calls on memory and never gives a single thought about it working effectively. So too, we take it for granted that when healthy and we direct a physical movement, it will occur. We just have an unspoken faith in those inner abilities. Parents and teachers require we learn, and when we answer their questions, we get self-confidence and then learn on our own. When we're asked where did our ability to think and physically act come from, almost all of us avow that something has been given to us by a power greater than ourselves. That means that each of us, by our words and deeds, expresses that the mature functioning of our nervous systems constitutes a religious faith."

"*Belief*—" Bonnie corrected, countering my use of 'faith' by substituting the word she wanted used.

I did a half-shrug, agreeing to disagree.

"I find I always have questions about the learning process and the senses." Bonnie did not elaborate.

"Bonnie, it's certainly taken a lotta conversations for you and me to agree on the ground rules for our communication. But, we've always proceeded on the agreement that in the beginning, God created the heavens and the earth. But, after that, we both fine tune our words on how we relate to the world. Then, ultimately, we both agree that mature individuals have independent thinking with free will."

"Eh—ah—all right—"

"For middle ground, I propose looking at a basic, easy-to-understand model of the brain. Bioclock science simply points out how a fundamental unit of information is processed from heartbeat to heartbeat and that unit can often be matched with involved brain regions. For example, a person gives himself or herself the self-directive '*I will remember the name of the street.*' That represents a cognitive task that starts in the control center in the left side of the brain with '*I will*', extends to the word-memory region in the right side of the brain to locate the word-for-it, the name of the street, and if the name is retrieved, it is communicated back to the control center to complete a typical curiosity bioclock."

Bonnie gave a how-do-you-know-all-this look. "I certainly can't imagine any way to ever determine the exact order of events in the brain—"

"Disruptions of regions because of trauma give us some information about brain function and then reasonable conclusions tell us more. My hypothesis is that, in the conscious individual, a neuromuscular task, for example, is aligned, prioritized, to be resolved before a simultaneously appearing cognitive task. Let's say the self-directive had been: *'I will cross the street'*. That is a neuromuscular task that also begins in the control center in the left side of the brain, extends over to the long-term memory region in the right side of the brain, but this time the communication continues onto the peripheral nervous system to complete the task.

"Certainly, an individual who understands these basic processes can more efficiently use neuromuscular and cognitive abilities. He, or she, knows physical activity lowers stress. They know why sufficient sleep is required for long-term memory to become efficient.

"Indeed, a person who knows and accepts the way memory works can enhance the learning process and thrive—Just examine the way nature, itself, aligns the memory process. On the left side of the brain, the control center is ruled over by the person's first name and the pronoun 'I'. On the right side, most verbal data is processed, including generalized self-references such as, for me, 'the child of Daly' or 'the Daly son'. "

Bonnie's jaw made a drawn-in and then let-out movement that told me a disagreement was brewing. Bonnie was an active, bright, attentive individual with a warm personality that was ready-to-discuss-the-subject and see-what-can-be-learned. These qualities were epitomized by the use of her worldly blue eyes: in response to someone speaking, she almost always made eye contact and then facial expressions that communicated her feelings—such as joyful friendship when in agreement or preparatory feistiness when not agreeing.

"Bonnie, let's just use the example of how we interpret a dream. When you told me about a dream you had, you told me: *'In the dream, I was reading this book—'* Now suppose we interpret the

'*I was reading*' part as 'The Evarts daughter was reading—'. That would mean you, in your use of the memory process during sleep, were represented in a generalized profile as a family member, 'the daughter'. You were the dreaming 'child' who was incorporating the parental theme of 'read' into your thought process for use when you awoke. The memory process in the dream sets down '*The family daughter was reading*'—or some other general term such as '*Mom's girl is reading*'. A generalized *self-reference* by the dreamer programs '*the parent*' into the memory process and the dreamer, upon awakening, goes and imitates the teaching set down by the parent.

"A person who knows himself or herself," I was trying to speak casually, "can be relaxed and self-contained and understand that curiosity is a natural process in which something that is seen is to be identified by retrieving 'a word for it'. This person knows that a curiosity bioclock occurs anytime 'a word that answers a question' is retrieved.

"Teachers can use knowledge of brain function to make children better learners, first, by getting them sufficient physical activity, second, getting them to learn a little basic material and third, by asking them questions. Perhaps the best contribution my model makes toward achieving faith in oneself is that it simply demonstrates, in the healthy person not using a major compensating process, that there is no 'left brain' versus 'right brain' but rather one, unified functioning brain. An individual who knows and accepts that brain power is a gift is an individual who naturally acquires humility and sees the universe as magnificent, orderly and inspiring—and that begets the faith that there is a Creator of the universe."

Bonnie took a long moment of silence and I waited for her evaluation. "Annie Sullivan, Helen Keller's teacher, has always been an inspiration to me—" she paused—"and it would be real interesting for me to do the research and see if she taught that the finding of 'a word for it' was a developmental step in learning or an innate ability. That would give us a basic understanding of the origin of knowledge."

The next day, while on a long walk, I was thinking about Bonnie's comments. Somehow, my walk took me to the one-room schoolhouse. It was set back from the road and centered in a grove of trees. It reminded me of a tiny traditional meetinghouse or miniature country church.

I stopped and blankly stared at this now unused, small building. You know, I told myself, just about everyone you've talked to about the school has had a story of how they were affected by the sound of its bell. I had never heard the call of its bell but I wanted desperately to hear it. I closed my eyes. You're anywhere out here in the immense quiet and the bell sounds. You come to full attention. You know you must respond to the bell. I opened my eyes and felt startled and fully summonsed. I looked left and right for the others responding. No one. How did I manage to so completely imagine that bell? Will I be belled again? Is this in my mind as real or unreal?

My mind went blank. My eyes squeezed shut. Strange, I'm being haunted by a sound that I have never heard.

The school loomed at me. I bowed my head and silently apologized to the farm families for the way I had prejudged the educational value of the one-room schoolhouse. Without evidence, I had judged that it was of minimum worth and now I was amazed that "its bell" had produced intelligent thinkers, numerous college graduates, and successful business people. Each morning, how had Mrs. Alberts begun with a group of students in grades one through eight, taught lessons to each grade beginning with the first grade and then had that method work out so successfully? I choose to reflect on that question.

Factors in the success of the one-room schoolhouse must have been a good teacher, a home where learning was encouraged, an environment promoting competition and rewards for students who read books assigned to higher grade levels than their own.

What if I taught in the one-room schoolhouse? How would I get started to have them benefit from a month-long, introductory project to the 24-hour bioclock? Bonnie had had spectacular success with her kindergarten students observing the growth each day of an amaryllis plant. Maybe it would be best if you just explained to the first graders that each day a plant has very different responses to daylight, dusk, darkness and dawn. The green part is the top and responds to daylight by producing sugar. The part that carries the sugar from the top to the roots is responding to dusk. The root end has the white part that responds to darkness by storing food. The part that carries food from the bottom to the top is responding to dawn.

If the students observed a favorite animal, such as the cat, how could they be taught that "Tabby" and her relatives have 24-hour bioclocks just as we do, but that the cat species, unlike us, long ago adjusted its 24-hour bioclock and most cats prefer to be asleep during daylight, preparing to awake at dusk, awake at night, and preparing to sleep during dawn?

My reflections on the influences of bioclocks led me to notice that dusk was upon the land. "Rourff—rourff—rourff—" was a dog's unhurried yet raspy barking that was drifting in from somewhere on the distant prairie—It's as if that animal has forgotten what started him or her to bark—Or is this dog just expressing a bay inherited from wild ancestors of the long ago? Or does this animal have an imagination that creates far off enemies who must be frightened away?

Upon my return to the farm, Sue, with her two children, Christopher, about age ten, and Lisa, about seven, were helping to prepare vegetables at one of the frequent family get-togethers.

"Christopher," I asked, "what is your one favorite vegetable out of all the vegetables?"

"I like cucumbers," he answered.

"Well, it is very easy to grow cucumbers," I enthusiastically explained, "all you have to do is just stick little seeds in the ground and then pour some water on them everyday. Then, just like you and Lisa, the plants grow each day and rest at night. Now, if each of you had a little garden that was all your own, would you like to grow cucumbers in your own garden?"

"I'd rather go fishing," Chris declared.

"Can I stay up later if I grow them?" Lisa asked.

Sue loved to laugh—and after she laughed at the results of my classroom, she spoke: "Maybe the cucumber doesn't grow by day and rest by night—my father believes he can hear them groaning as they grow during the night and each morning he can see that they're bigger."

I was pleased that someone had lightened the conversation and gotten everyone—except me—laughing and joking about burping cucumbers.

"The top, or green part, of a cucumber plant, can grow toward the sunlight and the bottom, or root end, can grow away from the top when no light is present—" I explained.

"Is that a frog or a cucumber I'm hearing?" Someone spoke in a disguised, husky voice.

I looked over and saw the speaker was Jacqueline, one of Garth and Betty's daughters.

My straight factual approach needed to be more casual: "I want to wave an endive, not a carrot, at you to draw you into my work. Simply stated, the growth habits of the endive present excellent features for teaching bioclock science—"

While I had been speaking, Claudia, another of Garth and

Betty's daughters, had begun speaking in a rather interruptive manner: "I wonder how many Americans consider the endive an elite menu item just because someone pronounces it 'on-div' and makes it reek of the upper crust?"

"Unlike the cucumber, the endive is rarely talked about and that excludes it as a garden variety vegetable," Donn Hanneman, Sue's husband, contended.

I couldn't believe my teaching tools were getting licked as they were being enunciated. "Please, let's not have a heated-up discussion of just the vegetables."

"Shakespeare would certainly have said, 'An endive by any accent would still taste like an endive.' " Rick, Claudia's husband, stated.

I gave a false smile and bowed my head, hoping to hide from the group that they were getting me into a stew. "I'm worried that if we ruthlessly grind up the endive, the children will not savor it with their tongues and cheeks."

Gil, Bonnie's husband, gave me a 'that's-ridiculous' giggle. "I think we just need to toss up more vegetables to get your plant biology to widen its horizons."

"Yes, I think you should expand your explanation into a salad of ideas that features the endive," Donn concurred.

"The endive very much contributes to a healthy diet that's good for vision and good for hearing and it is a learning aide," I said, blending subject matter and purpose.

"Yes," Bonnie sided with me, "Let's not take something with such a simple and elegant presentation and spread it into a tasteless hash—"

"I often get a little salty when I encounter a bland offering," Rick commented.

"If we refer to the 'on-dive' as the elite 'vegetable du jour'," Claudia explained, "I think everyone will want to try it."

I was growing sad. I could have been cutting an onion for the way my eyes were watering.

"I think what's needed here is a spicy dressing," Gil said to stir things up. "I could make up something with chili peppers—"

"Oh," Betty exclaimed, "Gil's chili always doubles as a dip."

"I think the old boys used to have a way of making cucumber wine—and we should have them have a go at the endive," Garth suggested.

I realized Garth would not have a still moment until he had his old boys making a mash that substituted the endive for the cucumber and I wanted to prevent him from boiling away my clarity of thought: "Let's keep the endive out of any hot water. It's bad enough to think of the poor cucumber mixing with these dregs of sarcasm—and going down the drink."

"Might call that 'sourcasm'," Jacqueline added.

"In a pinch, marinating or cooking the cucumber in enough red wine would substitute for cocktails," Sue announced.

"Perhaps, a sage offering would save the day," Gil continued his peppering.

"I hope we're not going to let this project mushroom out of control." Bonnie pleaded.

"If we carve serving instructions right on the cucumber, we could turn the vegetable right into the recipe," Sue summarized.

"Oh, I'd relish a dish like that," Jacqueline quipped.

"Yes," Bonnie agreed, "that would be a way for us to chip away at the cucumber."

"Ah, yes," Claudia waxed philosophically, "that would be a good way to get the cucumber to be cool—"

"Please." I held up my hand. "If it's simplicity you want, you don't need to skin the cucumber to have it serve as an educational device—"

"I think the problem is that we're finding the Jay of cooking is no joy of cooking," Jacqueline reasoned. "Perhaps, what you need is to find a saucy topping for your presentations."

"More saucy!" Claudia added in a hot cheer.

"More saucy! More saucy! More saucy!" A chorus began steaming things up.

"Please," I requested, trying to put the lid on that buildup. "The proper manner and decorum here should be not to sing while the poor cucumber is being chewed up."

"Jay, I see you're getting yourself into a pickle," Sue declared—"I suggest you just preserve the cucumber for a later time—"

Later that evening, Claudia, who was a teacher specializing in child development, sat down with me at the round table used for dining, games, and verbal adventures. "Jay, I'd be very interested in hearing why you think your terminology, *the daylight, dusk, darkness and dawn subdivisions*, should replace such traditional vocabulary as the *hemispheres, corpus callosum, cerebellum*, and *brain stem*."

"Just the opposite. Those terms are anatomical landmarks but no one other than myself has shown how they fit into a unified system of brain function. My model of the healthy brain does fit them into a unified system. My *Unified Brain Model* shows the nervous system is organized and functions in *four contiguous subdivisions*: *the daylight* is on the left, *the dusk* and *dawn* in central regions, and *the darkness* on the right. Each *subdivision* sequentially rotates into a time when it, subtly or distinctly, dominates the interactions of *the subdivisions*. The *daylight subdivision*—with some qualification for the left-handed individual—is the preeminent *subdivision* and contains the individual's control center with most incoming neural information going to it and most outgoing information going from it."

"*Rotates into domination*? I'm not familiar with that as a psychological concept—"

"It's a descriptive phrase for the *subdivision* that's initiating or influencing neural events to make the individual awake, drowsy, sleeping or awakening. On a 24-hour basis, the process is that *the dawn subdivision dominates* with the tasks of *activating* two major routes to *the daylight subdivision*. One of those routes includes a part of the brain stem. When those two tasks are completed, *the daylight subdivision dominates* with the individual functioning in the awake state. Sometime later, *the dusk subdivision*, to take over *domination*, has the tasks of *inactivating* the two major routes of *the daylight subdivision* that *the dawn subdivision* had earlier *activated*. When that *inactivation* is completed, *the darkness subdivision* is in domination and the individual is in the sleep state. Thereafter, the process repeats.

"The *rotation of dominance* is the first secret of understanding brain function: during every 24-hour period, every individual experiences a sequence of four general levels of consciousness, (1) being awake but not cogitating, (2) falling asleep, (3) sleeping but not dreaming and (4) wakening up. The second secret is to realize that during each of those general levels of consciousness, every *precise* neuromuscular, cognitive, or dream, event is achieved by use of a *heartbeat bioclock*. That *bioclock*, symbol *HB*, is each heartbeat giving a jolt to the brain that initiates a single, or singular, use of a circuit through *the four subdivisions* to achieve a task. It's as if an elite vehicle was sent on a specific route through the four quadrants of a city. Each heartbeat, sometimes aided by neurotransmitters, uniquely empowers an event. Thus, each HB is the ticking away of the 24-hour bioclock."

"I'd be very interested in what evidence you have for your conclusions," she said, maintaining her highly efficient and thorough approach to academic matters.

"Well. So many of my conclusions have derived from studying the traumas and deficits individuals have suffered that I always first thank and give credit to the many wonderful and generous individuals who have had injuries or dysfunctions yet gave me their most personal histories in hope that we can improve our understanding and treatment of brain disorders.

"Next, my work, *The Periodic Table of the Bioclocks*, includes a wide-range of evidence for the 24-hour bioclock as the organizing principle of human life. It shows that changes in blood chemistry, breathing, temperature, and behavior, which are measurements that can be verified by scientific documentation, confirm the existence of the human bioclocks. Overall, the Table uses observations, reason and logic, to become a comprehensive method for gathering evidence on what is going on in any species.

"Daly's *First Law of Biology* states: *Every species has a daylight, a dusk, a darkness and a dawn, subdivision that function sequentially as a bioclock.* That means every single form of life started with a genome based upon a sequence of responses to the rotation of the planet on its axis each day. That basic, universal law is both an anatomical map and an organizing principle: it sets down contiguous subdivisions and the sequence in which each subdivision normally functions.

"*The Species Extension* is: *The human species has each subdivision rotate into dominance and use the 1) instinct 2) neurological 3) curiosity 4) reflection and 5) monitoring, heartbeat bioclocks to produce every precise behavior, cognition and dream.* Each species has an Extension of the First Law to define it by the bioclocks unique to it."

"The 'heartbeat bioclock' is a working hypothesis within your overall theory of the 24-hour bioclock?"

"No and yes. The heartbeat bioclock does produce a physiological effect that jolts the brain from the left side to the right side and empowers all neural events. Its existence does allow functions unique to the human, thus answer many of the deepest questions of brain functioning. The heartbeat bioclock is a revolutionary acceleration of information processing within the 24-hour bioclock. Each HB produces a millisecond interaction of the subdivisions, a functioning *loop* that achieves a singular task. In the many changes our species has undergone, the HB has changed the time to complete some tasks from hours down to milliseconds.

"Each heartbeat bioclock is initiated by, or associated with, a particular region. Each is then usually identifiable as a loop through the subdivisions. Partial, joined, or extended, heartbeat bioclocks can also be accounted for. I use as an example the individual who is walking into an intersection, notices that a car is approaching, and stops walking. In that experience, visual information on the left side of the brain linked with auditory information on the right side of the brain and the individual stopped walking. This was a neural process proceeding sequentially through the daylight, dusk, darkness and dawn subdivisions: the individual saw something, linked that vision-related data to the verbal-related data identifying a 'car' and then coordinated a neuromuscular response to stop and wait, all within a few hundred milliseconds."

"I don't see how information from the senses is incorporated into your functioning model?"

"Good question. Vision aside, most data from the senses goes to the right side of the brain and is mediated by the auditory system and often put into a verbal language. For example, take the mechanisms of dreams. Even during sleep, the individual is receiving and assimilating some incoming data from the environment. This data is formed into heartbeat bioclocks. For example, if the sleeper is about to fall out of bed, he or she experiences a heartbeat bioclock that instantly brings on full consciousness and an adjustment preventing the fall. In less intense events, the sleeper is awakened just long enough to adjust for the problem—too much heat, cold, or an uncomfortable position—"

"Are you saying dreams are only assimilating sensory information?"

"No. Dreams are responding to both external and internal events. They deal with everything from environmental threats to their primary internal purpose of converging and aligning units of information for use during the awake state, a purpose mediated by *REM-dreams*. In sleep, the darkness subdivision is dominating and verbal information is converted into a dream and routed from the right side of the brain to the left for immediate or eventual processing. The actual theme around which a dream forms can often be hard for either the dreamer or analyst to identify. Humor is meant to be a most frequent, beneficial theme, but a person, upon awakening, may not find a dream humorous, but disturbing, if not frightening. It's the idea that a dream that has a lion whose head is that of a deer is delightful but a dream about a deer whose head is that of a lion could be frightening."

"How does your theory account for a species that has a life cycle under 24-hour hours?"

"The equation for any bioclock is the same as that of the 24-hour bioclock: $B = S1+S2+S3+S4$. B is the total time it takes to complete one rotation of any bioclock. Each S is the duration of time that *Subdivision* functioned in that rotation.

"A given species may have had modifications affecting its S developments that have made its 'rotation' faster or slower than that of the human species. For example, a one-cell species that takes one-hour for a rotation of its subdivisions has a rotation 24 times faster than the human rotation of dominance. If the one-cell's rotation results in replication, it shows the simultaneous organizing and fulfilling of a life cycle."

"Have you tried to get your work reviewed in any of the scientific journals?"

"I'm in the process. At present, the armies on the plains of

Boston believe all medical discoveries are made in the offices of the *New England Journal of Medicine* and I've asked them to challenge *NEJM* to bring forth an alternative model to mine that better explains the mechanisms of neural disorders. I suspect that the past models of the brain available to that journal contained so much patchwork that the editors have forgotten that the brain does have a system by which it functions. My *Unified Brain Model* seems to send them into shock. I give them my simple observation that each time an individual uses long-term memory, he or she technically experiences a moment of sleep. They answer with the mysterious statement that they only want to serve the MD— Yet these same people admit that they do not understand how memory works and why we sleep and dream.

"I am now communicating with the Mayo Clinic and other institutions, but for now, I'm just advising anyone involved in a major neurological workup to make sure my model is included in the process."

Our topic turned to autism and I asked Claudia for her views on the subject.

"I approach any overly active child with an emphasis on walking and exercising with that child and always thoroughly commanding his or her attention. Most hyperactive children are not autistic. The children who are distinctly afflicted with autism do need us to give them patient and devoted love and look into everything for the cause and the cure."

I gave her remark a long confirming silence.

"What do you think of all this talk about mercury and autism?" she inquired.

"Metals are in this thing from beginning to end. In *Bioclock* DNA theory, each of the strands builds the matrix for a side of the body, a feature that explains how a genetic disorder can result in the formation of only one of two 'mirror features', such as one eye or one ear. This science needs to study how mercury poisoning apparently can result in an individual missing an organ. In the baby and older individuals science has done a good job of showing that metal poisoning from lead is vicious—probably from blocking the electrical flow of negative ions—I personally call the way arsenic goes to the fastest working parts of the brain a series of mini-tumors—and I personally would like to find out if any child has a single, very high spike of body temperature before the onset of autistic symptoms. As you say, let's look into anything that can be a cause."

She spoke most sincerely and dedicatedly about her work with teenage individuals and some of the research that she was doing which was specifically involved with psychological problems unique to that age group.

She spoke of the courses she was teaching at the University of Minnesota and her plans to compile a book of her work.

We shared many stories about the dedicated teachers and researchers who had inspired our lives.

"Claudia, I do find your comments on all these subjects to be filled with insights. I do encourage you to get that writing done."

She checked her watch: "TBC," she announced.

It took me a moment to realize 'TBC' was part of the new shortened language that meant 'To be continued'. "TBC," I agreed.

I realized that if, in my punctuation, I used a dash to substitute for a period when a speaker's sentence wasn't distinctly declarative or didn't overwhelmingly reflect the speaker's body language, maybe the English language did need to better serve communication.

MONTHS LATER, I WAS AGAIN ABLE to rule my schedule and was on my way to the harvest fields. In a pleasant, relaxed mood, I enjoyed misty farm vistas through metronome windshield wipers. A car on a long arcing runway in each valley. Concrete, metal and salt-shaker silos. Occasional perpendicular side roads. Towns located by spindly-legged water towers and elevators. Remembered Garth wit and why's.

As always, the farm family gave me welcome.

I wanted immediately to continue the tradition that Garth and I shared of swapping "humor rumors" as Garth called them, "truer rumors" as I called them, or "rumors of humor" as Jacqueline called them: "If Lewis and Clark had made it to this region," I told Garth, "they would have named it 'Lake Minnesota', declaring it to be one of the Great Lakes, and describing it as 'The lake of ten thousand islands.' They also would have declared the Zumbro River unique because if you travel downstream on it for two days, you arrive back at the very spot where you started out."

"Too much water at one time is as bad as no water at all." Garth said most seriously. He went on to describe the disasters of floods, the destruction of crops from severe hail, and the bad timing of excessive rain during harvest time.

Almost the only time Garth did not compete with me by presenting his own humor rumors was when there was an overriding concern for the success of a farm season. Today he had no time for light-heartedness and it was so seldom that he displayed anxiety, anger, or bad mood—never mind all three at one time—that I went quickly and quietly to the sidelines.

Garth moved his head from side to side several times. That meant he was thinking about something he disagreed with.

"The media keeps talking about global warming," he continued, "but I don't hear them talking about how a small shift in the weather pattern could be enormously and permanently destructive to a large part of the black soil belt that forms the most productive part of the bread basket. The northern border of that line of black soil only runs about a hundred miles north of here and it wouldn't take much of a change in the weather patterns to devastate it."

I began to picture the line of that black earth as it extended east through one state after another. "You should superimpose," I told myself, "a black-soil map over a map of the United States—"

Garth made a head gesture toward me and held a stare at me: "Your study of the earliest farmers"—he pointed a finger at me—"should include how carefully they managed the supply of water—for drinking—for crops—for hygiene—for the mutual benefit of a community. They learned to dam water, store it, and use it in lean times. They learned how to mound up rows for crops and make irrigation ditches that were used for the frugal use of water. They learned to water thousands of plants at one time. They increased productivity many times over." Garth was taking time to reflect between his comments—almost as if in-between his statements he was thinking about personal memories. "Even then, the success of the harvest was a year to year thing."

I wondered if he was remembering the stories he had told me about farmers who had learned from the birds, the bees and the beavers. "Maybe," I pondered, "it's actually Garth who's setting down the history of farming and my writing is in large part just the use of his material—"

"Ultimately, those developments and the engineering of wells, water sources and channels, along with sewerage systems, did produce revolutionary developments in the history of civilization—" he summarized and then turned to look out the window.

Uncharacteristically, he promoted a long silence. "That distant cloud formation seems to be climbing—could be forming a thunderhead—" He seemed to be talking to himself more than to anyone else.

It was late-August. For the next ten days, it did not rain. A walk with Garth always leads to a talk, so to it we went. The corn plants looked a healthy green, but I asked, "Are they getting enough water?"

Garth broke off an ear and tore the silks down a few inches. "The kernels grow from the bottom of the ear to the top. This ear has kernels growing all the way to the top, which means it has been getting enough water. The development of the kernels also depends on a very tight schedule of pollination. Each kernel has to be pollinated by the male part, the tassel, and there is a brief period when that must be done or the ear won't be full. You can see that if every ear was missing a third of its kernels, the entire crop production would be down a third."

As always, a long conversation with Garth led me into an immersion in farm reflections: Usually, as each new generation of farm families grew in physical strength, more production occurred. The family members gave each other a helping hand and lent a hand when a neighbor was in need. Some members enhanced everyday life with handiwork that rejuvenated hand-me-downs. Occasionally, a hired hand was brought in to handle a range of chores. Business was conducted by handshake. Government directives were usually disliked because they were bossy, not evenhanded, and did not address the small increase in the price for crops the farmers had gotten over the past fifty years. Religion thrived but during a busy season was often expressed quickly and efficiently: I thought that the best, basic religious service anywhere occurred in Minnesota when a farmer pressed his hands together in front of his chest and mentioned, "I thank the Creator for providing these two hands which are all the tools I really need to provide for my family."

Almost always, the farmers were open to new tools that saved time-and-motion to reward them handsomely with more time for turning observations into evidence for theories that made for discoveries, inventions and better management.

In my visits, I was always pleased to find that so many individuals in the farm community were friendly, disciplined and curious. Some misguided adults and some children believe that if they don't get everything they want, no one else should have anything. I admired the way the farm community kept children from coveting, from being spoiled, and from becoming bored and whining about it. Indeed, I admired the family teamwork of the farmland and indeed how it produced "the heartland of spirit."

In relaxed discussions with the children about my work, I found them most helpful. They frequently offered insights and excellent innovations. They were good detectives in observing patterns for bioclocks and they amazed me with many practical ideas for observing, explaining and teaching bioclock science. One of their plant studies, in particular, was a delight: they introduced rulers to make daily measurements, wrote down their observations, including physical descriptions, and weighed the plant by placing it on the scale, container and all, each week.

One afternoon, Kirsten Gandenberger, one of Bonnie's and Gil's daughters, was explaining her school work to Garth, Betty, and me: "In science class, we learned that the human genome project was collecting more than two billion pieces of information about the DNA blueprint."

"Indeed, I confirmed, "but those billions of pieces are anatomical parts and you need an organizing principle to know how they get the job done. Since our instant responses and reflexes depend on whether we're awake or asleep, we can say that everyone is first 'organized' to go through general states of existence each day: namely, the awake state, the state of being drowsy, the state of being asleep and the state of awakening. If we agree that those four states are roughly related to the times each day we call *daylight, dusk, darkness and dawn*, we can align some of the organization.

"We often use the terms *daylight, dusk, darkness and dawn* just to mean the time of day that's associated with the type of light we're experiencing. But bioclock science includes all the events that produce each one of those four phases of the 24-hour earth-sun light cycle. That means each phase is really formed by all the geophysical forces affecting us and the planet during that phase. Can you name any of those forces?"

"Gravitational and electro-magnetic forces."

"Yes. Excellent. And what geographical effects do we get from daylight, dusk, darkness and dawn because of the local area in which we live?"

"*Effects* such as we organize more vitamin D from sunshine during the summer than during the winter," Kirsten stated.

"Yes, indeed, our day and season, winds, weather, the sea level we live at—all affect our biochemistry and contribute to our *response to daylight*, our *response to dusk*, our *response to darkness and* our *response to dawn*. These are *the four responses* that have a major influence on how we are organized. These are the four responses timed as, or based on, the 24-hour bioclock. You, me—everyone—from genome to adulthood—has his or her billions of parts organized to work for a 24-hour bioclock each day. The motto of bioclock science is: *Follow the fours*. I lifted four fingers up in front me in a gesture of confirmation: "*Four phases lead to four responses that require four subdivisions*."

"Four subdivisions?"

I tapped four fingers against my forehead. "The human brain needs four different regions to organize the four responses: region one *for being awake*, region two *for falling asleep*, region three *for sleeping* and region four *for awakening*. Each of the responses is different so each has to be coordinated in its own region of the brain. We call each of those regions *a subdivision*."

"Doesn't every little thing the brain does—need a subdivision of its own?"

"Well, the brain does need a different region to initiate each task. For example, it has its awake center, called S1, on the left—I pointed—and its sleep center, called S3, on the right—I pointed. A region rotates into dominance over the others to impose tasks on the others. Stated another way, each region coordinates the required neural and biochemical events to produce its general state of behavior."

We did a high-four. I smiled.

"How do you know what's occurring where—?"

"Such things as injuries to the brain tell us that tasks that can no longer be done probably had been getting done in that region. Anatomy and physiology tell us things. We number the four regions of the brain as the responding subdivisions: *S1, S2, S3 and S4*. We then document, or at least associate, a behavior with events occurring in a particular region during its special part of the 24-hour earth day. Thus, for example: when we are awake we initiate conscious decisions from S1 and the other subdivisions assist with tasks to carry out those decisions. We respond to the world through one general state at a time and our four general states, in effect, 'rotate' us through a 24-hour bioclock."

"Is a bioclock—like wound up and it operates?"

I held in my smile.

"A 'bioclock' is short for *a biological clock* and each bioclock is only a repeat pattern of behavior or biochemical activity. It all started when the first cell on the planet was organizing its genome in four subdivisions as it responded to the four phases. Those four subdivisions were the first *S1*, *S2*, *S3* and *S4*. A superb and simple bioclock is just a pattern of four subdivisions that function and replicate.

"When we do go looking for a pattern that might be a bioclock, it makes sense to start with *daylight* as the first phase. To begin with, all species give prime importance to the effect of light on biochemistry to produce sugar and energy for building. Even the most unusual species arises from that basic relationship. If we can locate a region in a species that is responding to daylight, we call that *S1*.

"In our species, we can follow a remarkable blueprint—a living blueprint—of S1 prenatal development with pre-eye and pre-vision structures that align for possible postnatal stimulation from light and the building of an individual's control center in the left side of the brain.

"Next, S2 is *the dusk subdivision* which mediates the organism's dominant response to dusk. It developed as a part that slowed the organism's physical activity as less light became available. S2 also becomes the connecting route of S1 to S3.

"S3 is *the darkness subdivision*. It started out as the part of the DNA with a unique response to darkness and it has now progressed into a developmental process that forms into the pre-hearing and pre-ear structures and then into the dominant ear and coordinating region of the brain for the sense of hearing.

"S4 is *the dawn subdivision* which mediates the organism's dominant response to dawn. It developed as a part that accelerated the organism's activity in response to an increased amount of light –

"Everyday, from the first day of the first cell onward to all the organisms extant today, one cell and multi-celled, the S1, S2, S3, and S4 subdivisions have been at work organizing biological functioning and reproduction.

"In summary, the organizing principle of every life-form is based on, or is a variation of, the 24-hour bioclock. An individual cell traces back to a 24-hour bioclock and maintains a 24-hour bioclock. Cells in a multi-cellular organism trace back to a functioning meshwork that formed into an overall bioclock by making adaptations to internal and external events such as changes in temperature. If a life-form is influenced by any other bioclocks, each bioclock only exists as a pattern of influence on the S1, S2, S3 and S4 of the overall bioclock."

I thought my explanation was over, but Kirsten had her cat-like surveillance.

I reached into my briefcase and removed a pamphlet. "This is a three page Study Guide to *The Periodic Table of the Bioclocks*, a document that lists bioclocks and how they work. This is your copy. I'd like you to look it over and get back to me with any questions or suggestions you have about it."

Kirsten went right into reading the copy. "I'd suggest you use a series of illustrations to help the reader understand how each bioclock is formed by interactions with the four subdivisions of the 24-hour bioclock. It would be helpful if there were diagrams of the brain that displayed the locations of the subdivisions and how they were interacting—"

"Oohw—" was my barely audible gasp. I smiled and nodded, rather stunned at the quality of an eighth grader's suggestions, each of which immediately earned my respect. I was amazed at her sense of organization, ability to edit, and willingness to comment on a formal document.

"All good ideas. Any particular suggestions for making my presentation less abstract and more down to earth?"

"Something hands on—" Kirsten reached over to the Lego box, removed two pieces and clicked them together.

"Excellent idea. Four of the core chemicals which code for genetic information are adenine, thymine, cytosine and guanine. Each Lego can represent a chemical that is one of those billions of pieces of information. The joined pair can represent a base pair of the DNA code.

"Let's see what happens if we go on to a far larger numerical representation. This time we'll represent one of the two DNA strands." I reached over and selected four pieces: one white, one green, one blue, and one red. "If we let white represent *the daylight subdivision*, green, *the dusk subdivision*, blue, *the darkness subdivision* and red, *the dawn subdivision*, we can join them into the sequence of four subdivisions that forms the dominant strand of the DNA. I joined the four pieces. In total, how many of the pieces of genomic information are represented by this strand of the DNA?"

"That would represent one-half of all those billions of pieces of information."

"Close enough," I nodded in confirmation.

"Now, let's get the second strand of the DNA—" I took four more pieces in those same colors and attached them alongside the first strand, but in reverse order of colors. "Here is our DNA at work. Two strands joined side by side. Eight pieces joined into a cell but ready to go on and build into two cells."

I waited a moment while she examined the model. "How do you know this?" she asked. "How is this derived?"

"Remember we're just tracing back the 24-hour bioclock, day by day, to its beginning. That process leads to new science. Before, the theory was that hydrogen bonds join the two strands in a series of connections and later, proteins—or something—unjoin the strands. Bioclock science just observes that the first strand is a *bioclock-magnet* that draws in and aligns the parts that will make another strand. When this second strand is completed, an immediate change of polarity occurs and the two strands begin repelling each other and become two separate strands. Just imagine the two strands as two magnets and see how they can be positioned into configurations that join, repel and separate."

I separated the strands. "Now each bio-magnet strand can go on to draw in the parts for its own second strand." I took eight more pieces and added four to each strand to complete the models of two cells.

"This process continues, just 'bioclocks' along, and produces more and more cells. In the case of the human development, a group of cells aligns its bioclocks into complex stages described as the *blastocyst*, *embryo*, and *fetus*, with each stage, of course, also seen as functioning in conjunction with a maternal bioclock. If functioning can continue, eventually there is an individual with the mature 24-hour bioclock of being awake, going to sleep, sleeping and awakening. It sounds complicated, but if you start with a basic biochemical pathway forming the four subdivisions, you can go on to the complex circuits of the 24-hour bioclock.

"That complex bioclock circuitry, by the way, accommodates a most flexible range of uses, such as when it adapts for a person to take a nap—a nap is a mini-24-hour bioclock, an accelerated neural trip through the four subdivisions. Thus, each time a person on a very hot afternoon has a siesta, he or she has simply experienced, neurologically speaking, an accelerated loop of the subdivisions."

"Most farmers think if they have time for a nap, they have a time when a chore can get done that otherwise wouldn't get done," Garth mentioned. "But, then," he opened wide his eyes to emphasize, "the farmer does first think about a chore that has to be done and then thinks about just what can be done not to have to do it."

"Oh, I know lotsa times," Betty commented, "Garth takes a nap when there's a chore to be done and he thinks he's so smart when he awakes and the chore is done."

The subject of conversation became sleep and dreams. Only on some occasions did the farm community agree with me on the meaning of dreams. Many of these individuals insisted that my interpretations were not valid in their cases. It did not matter to them that in a course at the Harvard Medical School, I had done a codification of dream content: I had examined more than five thousand dreams and found basic themes common to all but about one per cent of the dreams. However, two out of every three people in the farm community for whom I interpreted dreams went on to disagree with me and state that they had formed their own beliefs on the meaning of dreams. Many also presented their own contexts to back up their interpretations. Usually, by mutual agreement, we agreed discussion of the subject was to be continued.

In the fall, the summer social talk of "Wieners and beans and How about them Twins—" changes to "Lutefisk and lefsa and How about them Vikings—"

One event within the prairie autumn is a uniquely intense experience that thereafter becomes an all-pervasive 'generic' memory for the harvest. In that season, a visit to the grand and great American prairie consists of sunshine joining the plants at maturity to highlight the fields and unify them into an empire in golden glow. The traveler to these harvest fields has a most special, visual adventure that enriches into an emotion of admiration, attraction and endearment for the land. It is in that moment of being surrounded by this experience that the viewer's reality expands into a sixth sense of awe for "the plants turned gold."

This was that moment before the start of the harvest, a moment of majestic, spiritual immersion. In it, the year's achievement was appearing on a vast, outdoor stage filled with crisp, seemingly carved-out-of-wood plants and the number of these plants was just increasing, their range expanding outward to enlarge the boundaries of this open-air theatre. The visitor was climbing over one hill to the next to see more of it and continuously one vista was simply taking over from the one that preceded. Slowly, out of these squared and rectangular dwarf forest sections of yellows and browns, there was arising a unifying, clear, gold air as if a shine was being sent upward to the sun and the viewing of its radiance was reaching the eyes of the astonished.

I was there, then, at that one exact moment, when acre by acre, one kilometer after another, a mystical act was occurring wherein plants and fields were forming a simple and plain continuum. This union became a unique presentation of joy as the vertical plants went out over the horizontal fields and the two formed into one, calm and stable, immense harvest field that gave witness to the honesty and worth of all who contributed.

In this stillness, one believes all the contributors are somewhere nearby and one looks around for them, imagining that they must be near and formed into a brigade of a single line in which each contributor is handing along one food basket after another and each basket is destined for a different individual around the world.

Harvest is a hug with nature.

As awed visitors have experienced harvesting time in Minnesota and then traveled the world to tell of the experience, an around-the-globe legend has come into existence that "The fields in America are lined with gold."

There is *The Favorite Farm Memory*, a book that has only *One Thousand Words and One Picture* but strikingly and overwhelmingly communicates the theme of its story. Sometimes, though, remembering the book's precise words and picture cannot be achieved and there is only a white flash from memory. "I know the contents have something special to do with farm life that I want to remember—" The watery haze is over the eyes. "Why is it that being unable to recall a farm-life-moment can produce an inner-driving-feeling of emotional frustration? Was it a memory of enjoying a day's work? A memory of a special place of meditation? A delightful place to visit?" These questions bring a mood of acceptance. A feeling of love. A knowing that something was right. I smile: I wonder if, maybe, for all of us, thinking about farm life is one of those archetype things? Whatever, even remembering that there had been an experience that brought a most rewarding, meaningful mood is good. An attempt to have memory bring up a moment at the farm is a satisfactory adventure as it does bring on a warm, positive attitude about living and makes words ease into existence: "Sometimes, the greatness in all of this searching is simply to be grateful for the fact that while reading a book, you can live in that book."

Searching out the main street of a farm town on the weekend and finding *the local restaurant for breakfast* is great way to enjoy local people and quaint places. Karen Zeller, a friend from the farm community, graduate of the one-room school house and individual whose goal was to go and live in countries around the world while teaching English, joined me in a Zumbrota restaurant at a window table. Disappointingly, the view was snow in a horizontal flow.

Karen and I were discussing the range of responses to the seasonal bioclock in humans. I was detailing, "During the winter months, deprivation of light and greenery makes many people less bright and cheerful."

"It seems sometimes as if only those afflicted with severe Seasonal Affective Disorder know how devastating it can be. A victim can be severely depressed everyday for weeks at a time. I feel the ill effects of this—bioclock—should be given far more attention." Karen continued to stare out the window.

"There's certainly a very wide range of severity in the population affected by depression—we're in mutual agreement on how serious it can be for some victims—" I said, shaking my head three or four times. I did not like to speak of the ailment by its nickname 'S.A.D.' because I felt that routine reference often trivialized the ailment. I did not want to talk about how devastated I had been by the people who had been destroyed by the illness. I was momentarily unable to speak. I did not want to be pompous: "My research on depression begins with the way a certain clump of 'downer' words is handled by the brain in early sleep and intensified during certain seasons."

Karen gave me a quick look. "I encourage you to emphasize the importance of the seasonal bioclock and research ways to help anyone afflicted with severe depression," she most sincerely stated.

"Karen, when an individual stares at the bleakest landscape and cannot tell where the snow-covered land ends and the totally-clouded sky begins, he, or she, should think of green—green grass, green shrubs, and green trees—ever-so-slowly growing beneath the snow and then think of the cloud cover as a gray-white mist that is evaporating and beginning to leave a full, light-blue-green sky. That will make the horizon visible and present hope and the challenge of where to go for new adventures."

"I think it's good to have overall goals," Karen confirmed, "but the depressed person really needs resources available at the time it strikes."

Karen was a healthy, well-spoken, pretty brunette who had the special quality of bringing an intense, but appropriate, facial emotion to each of her ideas.

I sensed she did not believe that I understood that a person with severe S.A.D. could be critically depressed.

There is a nod that is half-smile and half-acknowledgment and I gave that to her. "Let me begin with the way I approach the subject and you feel free to bring in comments at any time. I absolutely believe that people on the edge of depression and people beginning to encounter memory problems should do my

Think and blink exercises.

"Most of us average a blink about every six seconds. People getting depressed and people encountering memory problems often go fifteen seconds or more between blinks. Most times when we blink we are allowing a nervous system cycle to speed through a loop of the 24-hour bioclock and that rewards us with a small infusion of well-being hormones and opens the gateway, the circuitry, to long-term memory.

"Preferably, in front of a mirror, people who feel they're not thriving should do a *Think and blink* exercise in which they think resolutely of something that must be done. This exercise should begin with a full, decisive sentence using the pronoun 'I' followed by a verb denoting action, and then name the object of the action. For example: 'I will go to the grocery store.' Or, 'I will now phone Mary.' And then, immediately, the individual must go and do the declared action."

Karen went back to staring at the window and I had the thought that I wasn't getting to the problem. "If the individual declares the 'I will now' part but cannot name or remember the goal or destination, he or she should not worry about it but should just momentarily close the eyes in a long, soft blink. That often allows the word or goal to be retrieved from memory. So, if the exact goal of the plan doesn't quickly come to mind, the individual should just blink and smile and relax by doing a routine task that requires some moving around."

Karen gave me a smile that was part cheerful, part acknowledgement, and part thank-you. "And if that exercise doesn't work, how many times do you suggest repeating it?"

"Every time a person encounters a memory lapse, the *Think and blink* should be done just once and then the person should do a quick smile, go on to something else and wait for the answer to pop into mind after awhile. The hope is that *Think and blink* will develop into a relaxed habit. Then, thinking and blinking and feeling good should be so natural that it's just a part of the personality to think and blink and smile."

"I do believe it's time good formats like yours should be utilized," Karen enthusiastically commented. She went on to make positive comments about my organization of words and ask questions about the success of those who had made use of *Think and blink*.

I did not want the conversation to end with the idea of only one approach. "On an everyday level," I added, "there are certainly other helpful practices. Everyone should have green things around to improve their well-being. In winter, a rite of renewal anyone should do is: *Do The Poet Tree*—that's four words. That event is begun by looking at your favorite tree or a picture of it or thinking about it. Write down your observations of the tree and what you like about it. Draw, copy, paint, or make a model of your tree and try it in jump-out-at-you green. It can, however, be the simplest, quickest sketch. Make a bioclock chart for the tree by listing and describing at least three of its bioclocks. Three bioclocks are easy to name and study—the 24-hour, seasonal and reproductive. Celebrating a moment of thankfulness for one tree, especially one you've visited in summer walks, will usually make you want to smile triumphantly. The benefits of a Do The Poet Tree, by the way, are almost always made very much more delightful if two people participate together.

"Thereafter, anytime you feel the grays and browns of cold times, close your eyes and enjoy that tree. Say something like, 'I will go now, lift my head high and walk to that tree and see it standing tall and coaxing the sun to warm it.' Or, 'Now will I close my eyes, imagine I'm on a vacation, and I walk till I stand in the cool shade beneath that tree.' Always great is to do the tree dance by standing, stretching the arms out to the side, rising up on your toes, slowly breathing in while stretching the arms upward and saying, 'I'm a branch lifting in the wind—' and then breathing out as the arms are eased down. Do a series of these pleasurable stretches and you should be immersed in a superb, bright, cheerful mood."

I had first met Jacqueline, a graduate of the one-room schoolhouse, and an attorney, in Boston. After we had become engaged, we had made that first trip to the family farm to visit with her relatives. Now, after my many farm visits over many years, I interviewed her: "You've been a history teacher and history enthusiast and I wonder, with that perspective, what is your evaluation of your childhood years at the farm?"

"The experience of growing up on a farm was great. It taught us a basic appreciation of family life, responsibilities, and values such as thriftiness and honesty. Because of my mother and father's love of books, it was beneficial in giving us a better understanding of world events, everyday events, and it improved our cognitive abilities. Chores kept us physically active and healthy—especially Sue, who loved to be up on a tractor or doing tough chores. Overall, living on the farm allowed us to develop confidence in ourselves and thrive as individuals. Imagine our father running a farm with four daughters and no sons and never once making us feel that he was disappointed in us.

"Many of the farm families take time each week to give thanks to The Creator for the ability to provide for themselves with a minimum of help from anyone else. Life on the farm was competitive for the four sisters, but we usually respected and praised each other's achievements. Perhaps, that taught us something about charity, understanding other people's needs, and to some degree, how to better understand our own strengths and weaknesses. Each member of the farm family really needs to be self-reliant to thrive. A blizzard on the prairie can isolate a family for a week."

I nodded. "Bioclock science establishes direct connections between self-understanding and self-achievement. For example, the moment before going to sleep is a most special time for setting goals and emphasizing determination to achieve them. The events occurring just before sleep have prioritized processing during sleep and therefore have elite potential.

"Success with one's goals is further helped by repeating those goals immediately upon awakening. The last things reflected upon before falling sleep are given priority processing during sleep and become aligned as the first material to come into consciousness when the individual awakes in the morning. Reflect upon goals before sleep, get them worked on during sleep, and, upon awakening in the morning, reinforce the goals with positive attitudes and feelings of well-being about them."

Perhaps, it was the conversation with Jacqueline that led me to arise in the darkness of the next morning and go outside in cheerful anticipation of seeing nature at its most beautiful best: the farm black and outlined in red by the pre-dawn rise. As always, that view made me a monk in the moment between meditation and prayer.

That moment was ended by a crow at the highest spot on the tallest tree shrieking a sound that seemed to be the word "Hawk." On every farm road, there was an alert crow who believed the power line was a tree branch. Perched on a wire, the crow was on guard, ready to yell out "Hawk!" to warn its friends if a hawk did happen by. On farm after farm, a crow like this one started the day on the top of a tall structure. That subject had come up as I was visiting with Garth's brother, Wilbur, his wife, Evie, and their children, spouses thereof, and grandchildren. "Do any of you know," I asked, "of a farm family that has made a pet of a crow, somehow providing it with food and drink so that it returned each morning?" They responded as a team with a theme: "In the family portrait for this occasion, absolutely no one will smile."

My attention went to the sounds of other birds. They sang and sprang. "Bored" is not a bird's word. Ear led eye and I discerned the location of one with especially attractive melodic song. Some bird sounds seem to be words I use. Certainly, some form of memory was in their vocalizations: they were displaying curiosity and communicating location.

I concentrated on two grand birds who seemed to be playing a high-pitched language game wherein one would shriek out the name "Oliber" and the other would ask back: "Oliber?"

Each place where there was no sound, I took time and looked carefully, hoping that I would become the first city person to see a Minnesota silverback raccoon easing about. Minnesota golden gophers were seen all over the state and Minnesota timber wolves were occasionally seen, but in all the unique stories that included the Minnesota wild, if you saw a Minnesota silverback raccoon, you earned a special reputation anywhere unique stories were told.

I next visited Garth in the days after he had retired from full-time farming. He was enthused, happy, and engrossed in new adventures. Now that he had left the farming to others, he was working as a real estate consultant and using his farming expertise to teach others to manage their farms and that included the counseling of farmers on the practicalities of growing fuel plants. He also was helping farmers who were in the difficult process of selling their farms. As always, Garth brought his smile that sent out, and shared, extra strength, with others. I now noticed that his smile brought out deep, rich, facial indentations and proof that some people's eyes do twinkle.

I watched him work away at his fulltime hobby—the ever-expanding flower and vegetable gardens. His cap had the design of the flying corn cob on it and when he removed it and used a handkerchief to wipe his brow, he had the farmer's forehead—the top of the forehead white and untanned in contrast to the dark tan of the rest of the face. His sand-blond hair was now lighter in color and lighter in fluffiness and he enjoyed the style of not combing his hair. His glasses needed cleaning—as they always did. I gave myself a what-can-you-do smile.

My attention turned to the tractor working in the nearby field. It was being driven by Stan Streiff, a friend and neighbor who, by agreement, was now working the farm.

"Now that you're not out on that tractor," I nodded toward it, "but renting out the land, what really stands out in your mind as you look back on your years of farming full-time?"

I thought that Garth would begin with the winter chores and speak again of the teeth-chattering cold and the pain-causing wind with its blasts and swirls of sand that always sprayed and hit the face with sting upon sting. Bitterly freezing cold that hurt, numbing the hands despite gloves and caused him, no matter how lightly he touched something, to be painfully aware of each individual bone used in the contact. It was cold that went through the clothes and caused, each time he moved, joint aches; the cold that slowed brain use and weakened muscles. It was gripping-dry-ice pain that was endured only by constantly thinking, "Get it done and get back inside."

Garth continued in a reflective stare downward that was an exceptionally long moment of reflection.

I sensed I was getting colder and colder and somehow being compressed. Was my shiver psychosomatic or was there a Farm Winter Ghost whose grip was pain? The thought of my own worst, cold brutal day of work came to mind: 'It took me an hour to warm my feet. What would it be like to suffer that day many times a year for years upon years?' I looked at Garth and thought again of the outdoor chores he had done in the coldest of cold weather—sometimes a few minutes of exposure had killed flesh—the worst of the worst when highest winds and lowest temperatures made him cry out an uhmf with each breath as his lungs were pierced by little icicles.

What would stand out in his farm experiences?

He ended his long, serious downward stare, looked at me and then by me, and stared out to a distant field: "One year, I was really tired and I fell asleep while cultivating—" he was speaking softly and slowly—"Drifted over and wiped out a hundred-foot swath of young plants."

Some parts of the working farm had, for the present, served their purpose. The once silver top of the silo was now tinged with a rusted red. A line of five or six weeds, each about a foot-high, stood against the barn doors and, in its silence, the barn, itself, communicated that it was no longer on duty. I listened for a single cow to give a single uhhmmmrrr but not one was to be heard. The granary that Garth and I agreed was just as well called "the grainery" was empty and totally uninviting. Inside the garage, the oxen yolk on the wall was a reminder that when machines failed, draft animals were hitched up and made to work. The tools, the spare parts, the containers of nuts and bolts, but most of all, the F20 and Oliver tractors sent into odd jobs or retirement by giant machines like the combine, now served as a humble museum of the family farm.

Outside, the chicken coop was weather-beaten and angled over as if ready to flatten down. "I'll bet that there is only one other person besides myself who feels sad about no more chickens," I concluded, "and that other person is the one person who liked feeding the chickens and did not mind cleaning the eggs. And she also felt good about the times she went to town and sold the extra eggs."

I walked over to where Garth was picking a basket of vegetables. He started on the onions and I remembered that this is the spot where I had first had instant success treating mosquito bites and bee bites by rubbing fresh-cut slices of onion on them.

We tasted slices from an onion. Then he handed up lettuce leafs.

"Needs rinsing," I said, tasting some dirt.

"If you were a farmer, you would have gotten used to eating a little dirt now and then when snacking on vegetables right out of the field. I've eaten a lot of dirt—down through the years—and my digestion has apparently thrived on it."

I nodded. I thought of the years upon years he had driven an open but hot and fumy tractor that stirred gritty dust up into his nostrils and made him inhale dust and swallow dirt—and it was in that heat that he was soaking his clothes in sweat and then having them baked dry by the simmering air in the wrath of the Minnesota midday summer sun.

I felt dirt in my mouth—my tongue going dry—yet my attempt to force up spit was blocked by the sensation of a thick and oily substance in my throat and that caused me to gasp and force a wrenching swallow. Some muscles on my face quivered.

I noticed the furrows on Garth's face and imagined that if I asked about breathing hazards that he would take many minutes to give me a long description of his more than fifty grueling years of wear and tear in the tough breathing of noxious gases, sprays and airborne irritants. I felt dirty, pain in my neck and about to faint.

I stepped away. I watched Garth move along, up one row, down the next. From time to time, he would stand up and look out over the larger farm area. I knew that, in the quickest glance possible, he used sky, clouds, wind, air moisture, ground vibration and an inner sense of barometric pressure to anticipate the weather.

I recognized that when I looked over the fields, I saw the plants that were sticking out of the ground and I saw the sky over them but he was seeing everything that I saw and he was also seeing all of the life in the layer of black earth under the plants and he was thinking of how he had loosened that soil so that the roots of the plants were able to ease through it and incorporate the necessary nutrients—including some manure that carefully had been spread out so as not to burn the roots of the plants. He was seeing water-retaining regions that he had turned over several extra times to get them to dry and he was thinking of the number of worms per square foot that he liked to have perturbing the soil and adding their nutrients. And he was thinking back over the beginning of the growing season and how he had walked out into the field and ripped out burdock by hand to begin his first official interaction with the land and growing season. In recent days, he had become most concerned with the water content of the corn.

Now, at a time of harvest, he was aside the corn plants, checking ears, and pleased with the size and weight of one, snapped it free. He yanked it open in a partial husk, lifted it to breathe in its essence of maturity and gave a look at its yellow and white, full, glistening rows of plump, rounded kernels. He bit into them, felt his teeth tearing the sweetcorn, heard the kernels' crackly crunch and tasted a sweet, wet, cool gush on the tongue. He puckered his lips, compressed his mouth to squeeze back a refreshing spray into the throat and felt a reflex swallow. And with a deep, relaxing, head-lifting drawing-in-of-air for a breath, he knew that the living soil, the living plant, and he, had all come together in mutual agreement.

Always, in preparing to leave the farm, I took the time to do a relaxed summary of my work. My invocation began with congratulating the Creator on getting the universe started.

How goes the crop of bioclocks? To answer that question, let me pretend I'm giving Garth an up-to-date summary of my work. When we talk over ideas, the exchange is relaxed, the language sincere and the conclusions cheerful. My goal is to show how the brain works and his goal seems to be how best to use that information. Garth is a kind, intelligent individual whose curiosity brings challenging, innovative views on the subject matter despite his not having had advanced college courses filled with technical terms and intricacies.

For a trillion days the perpetual rotation of daylight, dusk, darkness and dawn has interacted with all life. These four phases of the earth day have given us the developmental principle of all biological existence. The *four phases, the four responses, the four subdivisions*. A paraphrase of this Ultimate Law of Biology expresses that you can see a star and believe that life is found throughout the universe: **A star brings a planet *the four phases and a bioclock.***

The Periodic Table of the Bioclocks allows a new interpretation of chemistry's Periodic Table of the Elements. Atoms are not alive, but their physics appear to form a *four-phase*, expansion-contraction cycle that has strikingly similar dynamics to the cycle of daylight, dusk, darkness and dawn. There are *four phases* of the atom: full electrons, losing an electron, minimum electrons, and gaining an electron. The *four-phase* principle should be used for the dynamics of the smallest and largest events in the universe rather than just theorizing that those events are "random encounters." Science must promote a cause-and-effect theory able to include any random events and any events for which it does not have instruments to measure. We applaud Einstein for his energy equation and now we seek the unified theory by establishing the "speed of darkness" as stars create light between two vacuums, sink into the direction of their hottest flares and only seem to be moving away.

Those involved directly or indirectly in neural disorders need to use bioclock science. For example, clinicians doing brain scans need the comprehensive illustrations in *Human Bioclocks* that show how the brain works. Those illustrations will allow them to differentiate the neural activity that contributes directly to a behavior from the activity that is just preventing a behavior.

The everyday doctor knows that the healthy human brain processes information by using two interacting bioclocks: (1) *the 24-hour bioclock* composed of *the general cycle of consciousness*, and *(2) the Heartbeat Bioclock*, composed of *the precise cycle of consciousness*. In the 24-hour bioclock, S1, S2, S3 and S4 rotate through a time of dominance with each subdivision having one set of tasks when dominating and another set of tasks when not dominating. The brain circuitry of the 24-hour bioclock is a superhighway allowing a round trip but it has entrances and exits connecting to local roads, streets and pathways that are used for *precise* side-trip loops lasting milliseconds. The picture is the 24-hour loop allows thousands of high-speed "mini-loops" through the subdivisions.

Analyses of behaviors are distinctly improved by basing their "neural loops" on the fact that about 9 out of 10 individuals are right-handed.

To identify each precise route of a heartbeat bioclock, first list the subdivision in dominance, then the chronological order of the other subdivisions involved in the HB: for example, *the S1-S2-S3-S4 neuromuscular heartbeat bioclock* (or just *the 1-2-3-4 NB*).

S1 is in the left hemisphere and when it has rotated into dominance over the other subdivisions, the individual is fully conscious and using a control center to prioritize and utilize vision-related information. When S1 is not dominating, it is receiving and aligning dream information.

S2 begins at the left hippocampus, is a central region consisting of half the brain stem and half the cerebellum, and has connections of S1 to S3.

S3 is in the right hemisphere and serves S1. It is the processing center for hearing-related verbal information and has connections to S4. When S1 is dominating, S3 is the principle bank of "words" in long-term memory and does some "word processing" at the subconscious level. S3 dominates to run the sleep state.

S4 is the other half of the central region, has connections from S3 to S1, and extends into the peripheral nervous system.

Pain almost always must reach S1 before an individual is conscious of it. But, pain information from the peripheral nervous system doesn't always have to go on to S1 consciousness. Sensory information from the right side of the body usually travels through S4 to S3 to S2 to S1. If pain is controlled by S3 mediated devices such as hypnotism, shock response, or anesthetic, it can be vented from S3 to S4 rather than allowed on to S1. Neck or spinal injury must first be eliminated as the cause of a pain and then that pain sometimes can be interrupted by turning the eyes quickly to a side and blankly staring. This can cause a moment of S3 experience with a break in the pain.

The autistic individual has deficient processing of information at both the General Level and Precise Level. He or she has limited or faulty interactions within and between the subdivisions. He or she has poor, slow, interrupted, or missing heartbeat bioclocks (with related hormonal problems). Does one very high-temperature spike bring on these conditions?

Loss of consciousness and sometimes a seizure may occur if an individual's information processing fails in S2.

An individual afflicted with autism may have moments "locked" in S3, thus looking dazed or "blank." This is the "half-asleep look" because the eyes are open but the individual is not succeeding in standard responses to routine stimulation. This event should be made a test for autism.

In some individuals, autism produces the savant who has

extraordinary processing of verbal information in S3 that is channeled into S4 where extraordinary neuromuscular coordination occurs, such as spectacular piano playing.

The making or hearing of any sound has a totally special place in the brain. Thus many sounds made by the baby, excluding crying, should be imitated and spoken back. The child sees an image, tries to link that image with a sound, *a word-for-it*, and then say, or react to, that word. It isn't gibberish: there is a reason why a particular sound—and not some other sound—is made. We can help the learning process. Minimize stress by promoting one task at a time and have elite sleeping conditions in which the child feels secure with minimum, agreed-upon light, sound and temperature. Be active with the child. Overall, we need to stop metal poisoning: positive ions interfere with the flow of negative ions in the neurons.

Autism, Alzheimer's, Parkinson's and many other disorders are associated with malfunctions of the S1 to S2 to S3 neural route linking vision-related information to hearing-related information. This makes each afflicted individual, regardless of age, a possible case of noise poisoning with possible metal involvement. Noise is at war with us.

The use of memory goes with the unique human achievement of abstract thinking. In a way, every time an individual uses memory, he or she journeys back through "stored" developmental events to find or form a potential beneficial response and that trip could go back "millions of years to the beginning of life." And then and there bring forward a lesson. In this way, the traffic plan for a successful memory search uses a S1-S2-S3-S2-S1 route: *the individual* was conscious, had a vision-related "image" of a tree formed in S1, refined that data through S2 to S3 where it *resolved* into "a word for the image seen" (tree), and that data traveled back to S1. A total time of milliseconds but a time in which the event was shaped by features unique to each of the subdivisions it encountered. A later HB may reuse that neuronal route.

Alzheimer's appears to begin in S2 at the hippocampus (a major rotary for neural routes). A heartbeat bioclock is initiated in S1 and becomes snarled in S2 and does not make the connection to S3. Often, the first sign is an individual stating, "I'm going out—" but he or she cannot name an exact destination. That HB started in S1 but was not able *to cycle* through S2 to S3 for retrieving the name of the destination. Often this individual becomes frustrated and there is a buildup of brain "physical-action" chemicals that brings on severe agitation and, in anger, he or she dashes without destination, out the door. Noise and metal poisoning are associated with the dysfunction.

Many people are so worried about memory problems that they almost make "I'm coming down with Alzheimer's." a self-fulfilling prophesy. Less stress, please.

Alzheimer's is first and foremost a war on the sense of hearing. We should all avoid exposure to ear-damaging loud sounds and be aware that some sustained, destructive sound waves can be out of the level of normal hearing such as some from RF signals, compressor motors and failing ballasts. Deterioration from Alzheimer's is often seen as recent events disappear as the onion of memory is peeled back into earlier relations, addresses, events and religious training.

Noise and stress heat the brain. For me, severe noise is associated with the snarls in the S2 neuronal routes of memory. Sip, sip, sip to drink sufficient fluids. Always sip at the TV. Use, or imagine using, water to relax. Wash by alternately applying warm and cool water on the hands, wrists, face and back of the neck. Shower warm, then cool. A relaxed individual has the best use of memory, so *think-and-blink* with active-voice sentences with cross-referenced goals, such as, "I will go walking in the Mall of America, a good Minnesota exercise."

An obsessive-compulsive disorder results from a self-directed command, such as "Wash up!" This command is a HB similar to a hypnotic suggestion. It originates in S3 and is sent to S4. It is not stopped in S1 to let the individual review and control it but rather the "command" goes on to S4, thus it is obeyed. This dysfunction keeps happening over and over again. Commonly seen commands are "Sink" (Clean the sink) and "Seal" (Check that the doors and windows are locked'). Bioclock science uses two protocols for help. In a mild OC (or a mild Tourette's) case, the individual learns the steps to bring the "command" into full consciousness and control it. In a severe case or a case with multiple compulsions, a professional needs to oversee the protocol.

When practical, schizophrenics should be told that we all hear voices. That is, anyone can "hear in the mind" some unusual S3 "words" sent into S1. We simply must learn that the inner words spontaneously appearing in the mind can seem to be in someone else's voice or seem to be someone able to speak directly into the mind. We must recognize that voices appearing within us are an everyday, natural event. Do we not "quote" the words of others in the mind? An individual who believes someone else is actually able to communicate directly into the mind must be asked, "When have you ever seen that happen to any other person?" Anyone who misinterprets the source of information appearing within his or her mind may react bizarrely, become hypnotized, lose contact with reality, or have hallucinations.

The schizophrenic individual—and others—may experience S3 "words" sent to S1 and converted into a distorted visual image. Such faulty processing of information may result in the individual seeing 'ants moving on the rug' or 'faces dissolving'. Anyone having this happen should be told that everyone experiences bizzare images in their dreams. Anyone who has such an experience should try to recognize that it is caused by a faulty processing of information in which a dream process occurs during consciousness. An individual should try not to respond actively to any event that seems weird and wild but should try to treat it as a daydream that will pass. He or she should do this by making a category for it: "I'm categorizing this as a daydream because it is just too bizarre. I must just close my eyes and count to ten."

The individual who worries about his or her schizophrenic-like events should be informed that: "Everyone who maintains a daily routine and lets family and friends help, usually leads an independent life." They are also to be told that often good medicines and group therapy help. Everyone who needs help or wants to help should examine how a little change in the sound of a word can lead to a distorted visual image: in a dream, "Bambi" can appear as a frightening "Bamb-bee," a "Leon" appear as a "Lion," a beer drinker appear as a "bear" drinker, etc. A word distortion is often the mechanism behind a hallucination or uncontrolled behavior. In mental illness such as delirium tremens, drugs or alcohol may be contributing.

Study the entire body as the "full neural bioclock." That can save hours in distinguishing the functions of neurons, spinal chord, and peripheral nerves. For example, the incoming data for a high-pitched, continuous noise gets to S1 which responds with "energizing chemicals" for the directing of muscle movements and identification of the noise. If this individual displays head tremors (S4-involvement), the neural route (nerves and neurons) of the entire process can be examined for faulty links. Observations may show that: "Head tremors may be the first sign of Parkinson's." To find the cause of any ailment, base a hypothesis on evidence of faulty functioning in a subdivision. Thus: *Parkinson's may result from noise poisoning* that depletes S1-S4 resources.

Some processes can compensate for nervous system disorders. For example, a person with distinct motor and/or memory problems joins a sing-along or dance group and just the music, itself, can stimulate S3 linkages to produce physical activity, smiling and singing. Musical displays, natural smiles and many well-being hormones are mediated in S3.

One postulation from Bioclock Science is that the human genome is a bioclock in which one strand of the DNA is a blueprint for a left brain structure and right side of the body while the other strand is a blueprint for a right brain structure and left side of the body. The development and connecting of these 'blueprint routes' does depend on bioclocks organizing tissues and organs to form the mature body. The point is, once the first strand exists, progress each and every day thereafter requires an *S1*, *S2*, *S3 and S4* forming one overall bioclock.

At birth, the proceedings of the *dominant* and *recessive tasks* must form into *the rotation of dominance*. This strategic development will only occur if *the tasks* of the *dominant* level begin to incorporate the tasks of the *recessive* level as *cooperating interactions*. Until birth, the recessive level largely functions in a secondary status while connecting the RIGHT-side-of-the brain to the left-side-of-the-body.

An individual has the best potential for adapting to a major injury of the left side of the brain if she or he has not reached the age of eight. The right side of the brain has some mirror circuitry to the left side that assists the individual's *cooperating interactions*, such as the use of the non-dominant hand. This circuitry is usually available to become the control center if the primary (dominant) process is compromised during the first eight years of life. Thus, in some brain damage, the individual is able to adapt because the vision control center in the right side of the brain substitutes for the vision control center compromised on the left side.

When S1 dominates, neuronal "words" in S3 link to and engage motor routes to S4 and the right side of the body but when

S3 dominates to produce the sleep state; these routes are made "inactive" and that prevents their neuromuscular connections and motor tasks. These "word" alignments mean that there is a verbal component to most body language, most emotions and most out-of-body experiences.

Many mind-body interactions align processes in S3. For example, the REM-dream is the mechanism that lets some verbal information *control* some neuromuscular processes. In sleep ("the preparatory state"), S3 is dominating, and the verbal information of the REM-dream is routed from S3 via S2 to S1, where it is converted into a dream. Upon awakening, some of the information aligned during the dream is available for S1 to use in the S1-S2-S3-S4 direction for neuromuscular control.

Self-understanding improves when an individual keeps a log of his or her own Periodic Table. The *periodic numbers* are the four durations each day of the 24-hour bioclock. Thus, on December 21 in Boston Massachusetts, if the individual was awake for 16 hours, going to sleep for .5 hours, asleep for 7 hours and awakening for .5 hours, the *periodic numbers* would be *16+.5+7+.5 hours*. These numbers are different than the numbers for the 24-hour earth-sun light cycle on that day which round out to an individual experiencing 9 hours of daylight, 2 hours of dusk, 12 hours of darkness and 1 hour of dawn or 9+2+12+1 hours of the phases.

Self-understanding and sensitivity to others improves with a knowledge of one's heartbeat bioclocks. The frequency, speed and precision of eye movements, smiles, stares, speech—these are studied by first labeling the subdivision in dominance and then listing, in chronological order, the subdivisions used. For example, a 1-2-3-4 neuromuscular heartbeat bioclock (*"NB")*. Investigate where and why routine heartbeat bioclocks have been delayed or are missing.

Bioclocks have been continuous from the first generation to the present generation. Down through the ages, variations bringing changes have occurred in the subdivisions. One *HB* once isolated a specific "tree from the forest." But now the HB has added neurotransmitters to allow more than one HB to be energized between heartbeats.

The general states of behavior that one species demonstrates during a 24-hour bioclock must not be assumed to be in the same order as that used by other species. The human species, for example, largely rotates into *the general state* of being "awake" during daylight, but a bat species may largely have adapted its *rotation* and be "awake" in the darkness phase. Bioclocks tell what is going on in the brain of any animal.

Science is best proved with the most simple, comprehensive observations that can be tested and verified. To that end, the everyday doctor needs to study a behavior for the health of its achievement by identifying the extended chronology of the subdivisions used by the heartbeat bioclocks. Thus, when a person uses a curiosity bioclock and has found the word-for-it and then speaks that word, these joined HBs are seen as a 1-2-3-2-1-2-3-4 speech cycle. The human race is unique because the heartbeat bioclocks allow abstraction, reflection, reasoning, choice and communication.

The next day, Garth and Betty were hosting their Thanksgiving Day get-together open to family members and close friends. On this most special day, Garth and Betty were cheerful, pleased to continue the tradition for one more year of the farm welcoming everyone home. I was remembering years back to one of these gatherings and a talk with Garth. I had wanted to goad him into one of his entertaining and educational commentaries. I had asked him, "What makes you work?" He had answered, "I work to have shelter, clothing and enough food to share." For me, those words represented the essential philosophical, religious, social and personal Golden Theme.

For all, a return to the farm on any occasion was a major time of appreciation and renewal. My mood was making me evaluate my time with everyone who was present. These friends had been a great help in educating me about farm life and in helping me with bioclock science and I wanted to find a way to demonstrate to them that our time together would produce a worthwhile result.

I became very quiet and brooding because of my reflecting on the hints of future changes.

I smiled as I watched Claudia and Rick's boys, Tyler and Ryne, running a race while yelling and laughing. Finally, Ryne, the younger of the two, finished runner-up in their race and became unhappy.

That scene brought the memory to me of years ago when Christopher and Lisa were running around the tire-swing that hung under the yard's maple tree.

And Bonnie and Gil's children were runners, I remembered. Chad at age six always ran or walked fast and when he and Garth had been in a local race, Chad had spent the race gleefully pointing back at Garth and yelling to everyone, "I'm beating Grandpa." When Kaylee was age 5, she had jumped from the van and ran toward Garth while yelling, "Grandpa, Grandpa, take me for a ride on the hilly-hilly road." Garth had made a roller-coaster adventure for the children by taking them on a fast drive over a nearby, bumpy road that had many crests. I loved it when Kirsten had been running about hectically and suddenly stopped to try to touch a butterfly on the clothes basket and as it was flying away, she improvised a dance with it. I had run second to Greg in a race because he had taken an outrageous shortcut and in order to humble him, I had him agree to another race in which, while we ran, he had to imitate my impressions of the sounds made by various animals. I had imitated cows, sheep, goats, horses, wolves, chickens, crows and the whistles of birds.

As that race was ending, there were two versions of who won but no question that Greg did do the better vocal imitations of the animals. And Jaci had out-raced me while being unconcerned that she was wearing flip-flops.

I was astonished at how clearly these flashbacks of the children playing around the farm had been imprinted on my mind. These re-creations were intense and the clearest use ever of my memory. And there was an organizing theme: in full, rich, joyful detail, I was seeing each of them running.

In the afternoon, Jaci, the youngest of Garth and Betty's grandchildren, and a fourth grader, was reading books on mythology, history and science that I hadn't read until I was in junior high school.

It had always amazed me how many people growing up in the great farm lands consider the phrase "traveling to the harvest fields" to mean a visit of a few days each year during the harvest season and do not understand that many of us from all over the world and for hundreds of years have been drawn, if not driven, "to travel to the harvest fields" to find work, better our lives and fulfill dreams of starting farms of our own. Of course, for some of us, our adventures have been successful simply because we met kind, wonderful individuals who renewed our faith in human nature. I think of Garth, who on my first trip to the farm, told me that if I wanted to get serious about the business of farming that he would teach me and I appreciatively nodded while seeing the farm silo as a lighthouse with a beacon—something that would, indeed, keep attracting and guiding me to this land.

"Maybe someday, I suggested to Jaci, "you and I could write a story that would update the history, science and legends that started with the early farmers and travelers to this region."

"If I wrote about those subjects," she answered, "it would be in poetry."

I suspected that Jaci knew that Homer had already gotten in the first written words on those subjects and she wanted to get in the last word on him and I decided to see how long she would let me get in the last word on her. "Ah, yes," I said as if I had really thought about it, "great literature often results when people travel to the harvest fields of such places as Australia—China—India—the Middle East—Sicily—Argentina—and Minnesota."

Other Writings by J. Fabian Daly

Study Guide for the Periodic Table of the Bioclocks. *$1.00*

The Guide summarizes bioclock science, explains basic brain functioning and gives every student a method for doing original research on any species. The researcher lists a species, or member thereof, and observes any behavioral and biochemical changes that form the repeating pattern that forms a bioclock.

Periodic Table of the Bioclocks for Homo Sapiens. *$10.00.*

This is gives a basic description of human bioclocks.

Human Bioclocks. $35.00.

This text is a summary of human bioclocks. It has a series of eighty illustrations and explanations of bioclocks. Written in everyday language, it gives a step by step explanation of how the brain functions and how to make the best use of that information. The brain's neural routes used in pain, memory, mind-body interactions and other functions are depicted. Explanations of interrupted or deficient neural interactions associated with autism, Alzheimer's, and other ailments are shown. It can be used to understand what is happening in the brains of other species. (Includes the Periodic Table of the Bioclocks for Homo Sapiens.)

Check or money order to BioclockScience.com.
No s&h within the USA.
All prices are for 2011 direct from the Publisher.
Volume discounts.

Birchfield Press
PO Box 573, Accord MA 02018-0573
1-781-749-7563 or e-mail: Bioclocks@hotmail.com

Supporting BioclockScience.com advances scientific research.

Only Daly's bioclock science is able to become an online, quick resource for evaluating a behavior for its relationship to autism, Alzheimer's and other disorders. It should be a place to ask about a pattern of behavior, submit videos, describe an event and get up-to-date answers. Daly is the world's foremost expert on brain function, but other crucial support for the project is needed. Those wanting to help should contact J. Fabian Daly.

BioclockScience.com should also be a site where research findings can be sent and made available: each species is to be identified and listed by its bioclocks in the ***Periodic Table of the Bioclocks***. Planned is an on-line, up-dated text of Daly's ***Bioclocks***. This text explains how bioclocks form the organizing principle of every stage of development from DNA to mature individual. Daly's Theorem states: *Every individual in every species has four sequentially functioning subdivisions.* That information alone greatly advances biological research.

Judge the worth of bioclock science on the Study Guide to the Periodic Table of the Bioclocks. Is it, or is it not, logical and the most complete explanation of brain function? Some censors and critics are afraid to explore and believe only those with medical degrees should have the right to set down a theory on how the brain works. They should be told that we have a brain to think with and the brain does have a system by which it functions. Daly asks his readers to be patient with his critics. "Some have been asked to change a lifetime of work and beliefs. Some are silent so as not to encourage possible quackery. And the vicious and arrogant should be given minimum attention."

Daly's forthcoming books deal with the subject of abortion, factors determining straight or gay lifestyle, media egos, and the Boston Public Garden. So much crass copying of Daly's work is occurring that the reader is advised to verify information sources.

Page 1 of *The Last Catholic*, J. Fabian Daly's novel of a Southie student in search of ultimate truth at BC.

As the streetcar reached the top of a hill in Brookline, I looked back at the downtown business buildings where Beacon Street started and from where it passed the State House, Common, and historical homes of the Back Bay, and eased westward across Boston like a ribbon of greenery with ivy. I knew just how old Beacon Street was. It had spent its childhood as a Native American trail, its youth as a British Colonial lane and was now, in its adulthood, an American thoroughfare.

This morning, riding out to Boston College for freshman registration, I was traveling the Beacon Street trolley car line. I was nervously thrilled and was attempting to concentrate my thoughts on the passing scenery, but the single question of what it would be like in college kept drifting in and out of my mind.

Finally, at a turn beyond Brighton where Beacon Street framed a view of a lake nestled in a hilly arborway, I alighted from the vehicle and started walking across this rural valley of the city. At a distant rim of it, I saw the Gothic tower of Gasson Hall with its four gray minarets rising against the sky like closed umbrellas standing in cynical wariness of even the clearest weather.

–1–

Reviews of *The Last Catholic:*

"The whole book made me sad to think that maybe I should have attended Boston College instead of Columbia."
—Jack Kerouac

"Thank you for allowing me to attend Boston College."
—Ms. Sullivan

"For serving as my guide, as Virgil once did, through the highways and byways of our charming city and revealing its humanity and its foibles, many thanks."
—Frank Ahearn

"The descriptions of the Southie shoreline should be made into a poem."
—Joan Buckley

The Last Catholic. $15.00
check or money order made out to Birchfield Press. No postage/handling fees for USA orders direct from the Publisher.

About The Author

From 1974 to 1983, JF Daly worked in the classrooms of the Harvard Medical School as an AV technician, eventually taking courses and solving fundamental medical questions. In those nine years, he spent up to twelve hours a day in seminars, classrooms and libraries with the sole purpose of learning how the brain functions. His discovery of *The Periodic Table of the Bioclocks*, bioclock science and the Basic Laws of Biology has led to a Rosetta stone of science that answers many of the fundamental questions of the earth sciences, biology and psychology. A 1976 graduate of the University of Massachusetts, he also did course work at BC, BU, Harvard and URI. He has worked for the Boston Parks Department as a gardener, plant biologist, and internationally known expert in roses. He grew up in a family of thirteen children in South Boston and went to public schools, including Boston Latin and South Boston High. He is a US Navy veteran. His many articles and books have established his reputation as one of America's most independent writers and researchers. Many of his books have been listed by his readers as among the best books they have ever read. He now resides at Linden Ponds, Hingham Massachusetts where he is compiling his basic cell theory: "To understand cancer, genetics must be combined with the laws governing cell-to-cell contacts."

Garth and Betty Evarts

Daly's PERIODIC TABLE OF THE BIOCLOCKS

(*PTB*) ***Study Guide***

Bioclock
(Symbol B)

Column listing each bioclock by *periodicity* (the time required for one revolution of the bioclock.)

Examples:

the 24-hour bioclock
the 365-day "seasonal" bioclock.

The Four Subdivisions
(Symbols S1, S2, S3 and S4)

Every revolution of a bioclock has four subdivisions that sequentially dominate function (*Daly's Theorem*). The length of time a subdivision dominates is its *periodic number*. A subject's *24-hour B* on 6/1/10 had these *4 periodic numbers*:

↙ S4: 45-minutes awakening ↖
S1: 15.5 hrs. awake   S3: 7 hrs. asleep
↘ S2: 45-minutes drowsy ↗

Observations and Measurements *(O&M)*

Homo sapiens

List the species or subject to be observed. The month, day, year _____. Age__. Dominant hand___. Gender__. Location (city) __. All other O&M_____.

Bioclocks is the study of biological patterns.

24-hour bioclock: *origin*

Daylight, dusk, darkness and dawn are *the 4 phases* of the planet's 24-hour encounter with geophysics (gravity, light, rotation on its axis, magnetic activity, geography, etc.).

All species function by using a bioclock based on *the 4 phases*.

Four Phases Lead To Four Responses

***The 4 phases* influenced every organism to function each day in a bioclock of four sequential responses:**

↙ S4. The dawn response. ↖
S1. The daylight response.   S3. The darkness response.
↘ S2. The dusk response. ↗

Trace a 24-hour bioclock back, day by day, to the first cell. That cell was driven by S1 sensitivity to sunlight and replicated as a *S1-S2-S3-S4 B* to become the blueprint and organizing route for al life forms. In a multi-cellular organism, a cell has its own bioclock and is a member of its group's overall bioclock.

The 24-hour bioclock

In the human brain, t*he daylight, dusk, darkness and dawn responses* have developed into *4 sequentially functioning, contiguous subdivisions*. Each subdivision rotates into dominance over the others to impose its own *general type* of behavior (awake, drowsy, asleep or awakening). This neural loop allows a 24-hour round-trip.

Genetics and conditioning affect behavior such that no one is perfectly in sync with the perpetual rotation of daylight, dusk, darkness and dawn.

In the human brain, all processing of information is prioritized to go from S1, to S1, or serve, S1.

Daly's Theorem: B = S1+S2+S3+S4

***The 4 subdivisions* of the human brain are linked to form the first level of information processing (S1 vision-data goes S2 to S3 hearing-data) in everyone:**

↙ S4. The dawn subdivision is in a medial region. ↖
S1. The daylight subdivision is in the left hemisphere.   S3. The darkness subdivision is in the right hemisphere.
↘ S2. The dusk subdivision is in a medial region. ↗

***The subdivisions* function in *"a rotation of dominance"* that produces *the four general types of behavior* ("the four general states of consciousness"):**

↙ S4. Awakening ↖
S1. Awake   S3. Asleep
↘ S2. Drowsy ↗

The brain functions by using two interacting bioclocks: the 24-hour bioclock empowers *the heartbeat bioclock*. Each heartbeat gives the brain a left-to-right jolt that results in a specific neural acceleration of information processing over circuitry of the 24-hour bioclock. Compare it to super-highway traffic that allows vehicles to make high-speed local loops

Depending on which subdivision is dominating, each heartbeat results in one or more *precise* (specific) neuromuscular, cognitive or dream events. These *heartbeat bioclocks* use S3 in their loops, thus the so-called conscious individual actually experiences thousands of millisecond episodes in the sleep state.

In the adult human, with some qualification for the left-handed individual, brain wiring for the 24-hour bioclock is:

↙ S4 coordinates right-side-to-left-side brain circuitry and ultimately brings on the awake state by *activating* the two neural circuits *inactivated* by S2. ↖

S1 in the left hemisphere coordinates vision and overall control.

S3 in the right hemisphere coordinates hearing and sleep.

↘ S2 coordinates left-side-to-right-side brain circuitry and ultimately *inactivates* two neural circuits to bring on the sleep state. ↗

Daly's PERIODIC TABLE OF THE BIOCLOCKS

hronological Bioclocks	The Four Subdivisions	O & M
olumn listing 9 species-wide ioclocks that are studied one 24-our B at a time or as a stage of evelopment: enomic ays 2-6 prenatal th day prenatal 1-day embryonic 50-day fetal st day postnatal elf-knowledge uberty dult	**The genomic bioclock is continuing life generated by the parents and consists of a dominant and a recessive strand. Prenatal bioclocks are preparatory, using S3 in dominance ("REM-sleep") to align neural routes for memory and mind-body tasks. Postnatal bioclocks allow S1 dominance.** **In the adult, S1 dominates for conscious control that uses a dominant eye/hand/leg:** **S4 aligns S1 motor plan for the body** **S1 is the "I", *the self* by first name** **S3 makes verbal data for S1** **S2 aligns S1 incoming/outgoing data**	The neural map of any behavior lists its *general and precise* parts. In S1 dominance (*the awake state)*, visual data first links to hearing data ("eyes link to ears") as the brain sees "the forest" and hears "din." Then, the dominant retina in S1 can use an "image" of one specific tree and link it thru S2 to precise hearing data in S3 to find "tree," the identifying "word" or "sound." (*Memory* is an internalization of this process.)
ioclocks within Bioclocks olumn for listing a bioclock by a onsecutive number of days within ne continuum of chronological ioclocks. Examples: *he 28-day lunar bioclock* *ne 365-day seasonal bioclock*	**The 28-day lunar bioclock:** **S4. waxing moon** **S1. full moon** **S3. new moon** **S2. waning moon** **The 365-day seasonal bioclock:** **S4. spring** **S1. summer** **S3. winter** **S2. autumn**	These bioclocks function by making changes each day in each of the 4 *subdivisions* of the 24-hour bioclock. The subdivisions have their influences each day on the pattern of behaviors because of changes in hormonal levels, amount of exposure to sunlight, moonlight and artificial light and gravitational effects on the nervous system.
eartbeat Bioclock (HB) olumn for listing the 5 types of *eartbeat bioclock.* *) the instinct B* *2) the neuromuscular B* *3) the curiosity B* *4) the reflection B* *5) the monitoring B*	***The* 5 types of *HB* are a neural hierarchy that functions from (1) to (5) when the person is awake and from (5) to (1) when asleep. Each *HB* resolves on the basis of amplitude and a prioritized neural alignment. Each *HB* has an identifiable *loop* through *the subdivisions,* as in *the S1-S2-S3-S4 neuromuscular B.***	The heartbeat combines with neuro-transmitters to generate information processing in *the subdivisions.* Disruptions of the vision-to-auditory or auditory-to-vision neural routes by noise and metal poisoning are associated with autism, Alzheimer's, and Parkinson's.

SOURCES FOR BIOCLOCK SCIENCE

OCLOCK FARM **© 2010 by J. Fabian Daly.** $20.00. In this novel, JF Daly visits a working farm in Minnesota. The work, dignity and iievements of farmer Garth Evarts and family are shown. This is a superb close-up of farm life along with the basics of bioclock ence. ISBN 0-9702401-1-2

MAN BIOCLOCKS **© 2010 by J. Fabian Daly.** $30.00. In everyday language, this text shows that the brain works by using bioclocks d how to use the PTB for self-improvement. ISBN 0-9702401-0-4

clockScience.com
nail: bioclocks@hotmail.com
. 1-781-749-7563

Birchfield Press
P.O. Box 573
Accord MA 02018-0573

Prices USA 2010.
Free S&H. Volume discounts.
Checks payable to BioclockScience.com

out the author: JF Daly grew up in a family of 13 children in South Boston. He is a US Navy veteran and 1976 graduate of U Mass ston. He worked for 9 years at the Harvard Medical School, took a few courses but left to concentrate on medical mysteries. He rked as an expert in roses for the City of Boston. His writing has lead to his reputation as a profound thinker and research genius. now resides at Linden Ponds, Hingham MA while researching how cells work: "To understand cancer, genetics must be combined h the laws governing cell-to-cell contacts."